U.S.-Mexican Security Cooperation: The Mérida Initiative and Beyond

Clare Ribando Seelke
Specialist in Latin American Affairs

Kristin Finklea
Specialist in Domestic Security

April 8, 2014

Congressional Research Service
7-5700
www.crs.gov
R41349

CRS Report for Congress ———————————————————
Prepared for Members and Committees of Congress

Summary

Violence perpetrated by drug trafficking organizations and other criminal groups continues to threaten citizen security and governance in some parts of Mexico, a country with which the United States shares a nearly 2,000 mile border and $500 billion in annual trade. Although the violence in Mexico has generally declined since late 2011, analysts estimate that it may have claimed more than 70,000 lives between December 2006 and December 2013. Supporting Mexico's efforts to reform its criminal justice system is widely regarded as crucial for combating criminality, strengthening the rule of law, and better protecting citizen security in the country.

U.S.-Mexican security cooperation increased significantly as a result of the development and implementation of the Mérida Initiative, a counterdrug and anticrime assistance package for Mexico and Central America first funded in FY2008. Whereas U.S. assistance initially focused on training and equipping Mexican counterdrug forces, it now places more emphasis on addressing the weak institutions and underlying societal problems that have allowed the drug trade to flourish in Mexico. The Mérida strategy now focuses on (1) disrupting organized criminal groups, (2) institutionalizing the rule of law, (3) creating a 21st century border, and (4) building strong and resilient communities. As part of the Mérida Initiative, the Mexican government pledged to intensify its anticrime efforts and the U.S. government pledged to address drug demand and the illicit trafficking of firearms and bulk currency to Mexico.

Inaugurated in December 2012, Mexican President Enrique Peña Nieto has continued U.S.-Mexican security cooperation, albeit with a shift in focus toward reducing violent crime in Mexico. The Interior Ministry is now a primary entity through which Mérida training and equipment requests are coordinated and intelligence is channeled. The Mexican government has requested increased assistance for judicial reform and prevention efforts, but limited U.S. involvement in some law enforcement and intelligence operations. Despite those restrictions, U.S. intelligence helped Mexican marines arrest the leader of the Los Zetas criminal organization, Miguel Angel Treviño Morales in July 2013. U.S. surveillance equipment, intelligence, and law enforcement agents also helped the Mexican marines find and arrest Joaquín "El Chapo" Guzmán—the world's most wanted drug trafficker—in February 2014.

The 113th Congress is likely to continue funding and overseeing the Mérida Initiative and related domestic initiatives. From FY2008 to FY2014, Congress appropriated about $2.4 billion in Mérida Initiative assistance for Mexico, including some $194.2 million provided in the FY2014 Consolidated Appropriations Act (P.L. 113-76). As of March 2014, more than $1.2 billion of Mérida Initiative assistance had been delivered. The Obama Administration asked for $115 million for Mérida in its FY2015 budget request.

Possible questions for oversight may include the following. 1) What have been the results of the Mérida Initiative thus far? 2) How is the State Department measuring the efficacy of Mérida programs? 3) How have Mérida programs been affected by the Peña Nieto government's new security strategy and how is coordination advancing? 4) To what extent is the Mexican government moving judicial and police reform efforts forward, and how is U.S. assistance supporting those reforms? 5) Are Mérida funded programs helping the Mexican government respond to new challenges such as the rise of civilian self-defense groups? 5) Is Mexico meeting the human rights conditions placed on Mérida Initiative funding?

Contents

Introduction 1
Concerns about Violence in Mexico 2
    Drug Trafficking, Organized Crime, and Violence in Mexico 2
    Drug Trafficking-Related Violence in the United States 4
Development and Implementation of the Mérida Initiative 5
    Evolution of U.S.-Mexican Counterdrug Cooperation 5
    Developing Cooperation through the Mérida Initiative 6
    Funding the Mérida Initiative 7
    Implementation 8
    U.S. Efforts to Complement the Mérida Initiative 9
The Peña Nieto Administration's Security Strategy and the Mérida Initiative 9
The Four Pillars of the Mérida Initiative 13
    Pillar One: Disrupting the Operational Capacity of Organized Crime 13
    Pillar Two: Institutionalizing Reforms to Sustain the Rule of Law and Respect for
        Human Rights in Mexico 14
        Reforming the Police 15
        Reforming the Judicial and Penal Systems 17
    Pillar Three: Creating a "21st Century Border" 19
        Northbound and Southbound Inspections 20
        Preventing Border Enforcement Corruption 21
        Mexico's Southern Borders 22
    Pillar Four: Building Strong and Resilient Communities 22
Issues 25
    Measuring the Success of the Mérida Initiative 25
    Extraditions 26
    Drug Production and Interdiction in Mexico 27
    Human Rights Concerns and Conditions on Mérida Initiative Funding 29
    Role of the U.S. Department Of Defense in Mexico 32
    Balancing Assistance to Mexico with Support for Southwest Border Initiatives 33
    Integrating Counterdrug Programs in the Western Hemisphere 34
Outlook 34

Figures

Figure 1. Map of Mexico 4
Figure 2. Individuals Extradited from Mexico to the United States 27

Tables

Table 1. FY2008–FY2015 Mérida Funding for Mexico 7
Table A-1. U.S. Assistance to Mexico by Account, FY2007-FY2014 36

Appendixes

Appendix A. U.S. Assistance to Mexico 36

Appendix B. U.S. Domestic Efforts to Complement the Mérida Initiative 37

Appendix C. Selected U.S.—Mexican Law Enforcement Partnerships 42

Contacts

Author Contact Information 44

Acknowledgments 44

Introduction

For several years, violence and crime perpetrated by warring criminal organizations has threatened citizen security and governance in parts of Mexico and presented serious challenges to the country's justice sector institutions. While the illicit drug trade has been prevalent in Mexico for decades, an increasing number of drug trafficking organizations (DTOs) are fighting for control of smuggling routes into the United States and resisting government efforts against them. This violence resulted in more than 60,000 deaths in Mexico between December 2006 and November 2012.[1] An additional 10,000 organized crime-related deaths likely occurred during the first year (December 2012-December 2013) of the Enrique Peña Nieto Administration.[2] In 2013, kidnappings reportedly increased by some 25%.[3]

Violence in northern Mexico and the potential threat of spillover violence along the Southwest border have focused congressional concern on the efficacy of the Mérida Initiative and related domestic efforts in both countries. Between FY2008 and FY2014, Congress appropriated roughly $2.4 billion for Mérida Initiative programs in Mexico (see **Table 1**). Of that total, more than $1.2 billion worth of training, equipment, and technical assistance has been provided to Mexico (see "Implementation"). Between 2008 and 2014, Mexico has invested some $68.3 billion of its own resources on security and public safety.[4] While bilateral efforts have yielded some positive results, the apparent weakness of Mexico's criminal justice system seems to have limited the effectiveness of anti-crime efforts.

Mexican President Enrique Peña Nieto of the Institutional Revolutionary Party (PRI) took office in December 2012 vowing to reduce violence in Mexico and adjust the current U.S.-Mexican security strategy. While Mexico's public relations approach to security issues has changed, many analysts maintain that Peña Nieto has quietly adopted an operational approach similar to that of former President Felipe Calderón, a strategy that includes close cooperation with the United States.[5] Congress may analyze how cooperation is advancing, how Mérida and related funds have been used, and the degree to which U.S.-funded programs in Mexico complement other U.S. counterdrug and border security efforts. Compliance with Merida's human rights conditions is likely to be monitored to ensure that anticrime efforts are carried out in a way that respects the rule of law. The extent to which greater cooperation leads to an increase in extraditions, particularly of high-level traffickers such as Joaquín "El Chapo" Guzmán, is also of interest.[6] Oversight of U.S. domestic pledges under the Mérida Initiative may also continue, particularly those aimed at reducing weapons trafficking. Congress could also explore how the use of newer

---

[1] This figure is an estimate. Cory Molzahn, Octavio Rodriguez Ferreira, and David A. Shirk, *Drug Violence in Mexico: Data and Analysis Through 2012*, Trans-Border Institute (TBI), February 2013. Hereinafter: TBI, February 2013

[2] Calculation by Milennio, as cited by Justice in Mexico *News Monitor*, vol. 9, no. 1, January 2014.

[3] "Suben Extorsión y Secuestro en Primer Año de EPN,"*Animal Político*, December 2, 2013.

[4] Mexico's security budget totaled roughly $10.2 billion in 2013 and $11.5 billion in 2014. It includes funds allotted for the Judiciary; Interior Ministry, Ministry of Defense, Attorney General's Office, the National Human Rights Commission, and transfers to the states and the municipalities in each state. Government of Mexico, "Mexico's Fight for Security: Strategy and Main Achievements," June 2011. Marciel Reyes Tepach, *El Presupuesto Público Federal para la Función Seguridad Pública, 2012-2013* and *2013-2014*, Cámara de Diputados, March and December 2013.

[5] Joshua Partlow and Sari Horwitz, "U.S. and Mexican Authorities Detail Coordinated Effort to Capture Drug Lord," *Washington Post*, February 23, 2014.

[6] U.S. Congress, House Committee on Homeland Security, *Taking Down the Cartels: Examining United States-Mexican Cooperation*, 113th Cong., 2nd sess., April 2, 2014.

---

tools—such as aerial drones that monitor criminal activity in the border region—might bolster current security cooperation efforts.

This report provides a framework for examining the current status and future prospects for U.S.-Mexican security cooperation. It begins with a brief discussion of the scope of the threat that drug trafficking and related crime and violence now pose to Mexico and the United States, followed by an analysis of the evolution of the Mérida Initiative. The report then provides an overview of the Peña Nieto government's security strategy and how it is affecting the Mérida Initiative. The report then delves deeper into key aspects of the current U.S.-Mexican security strategy and concludes by raising policy issues that may affect bilateral efforts.

Concerns about Violence in Mexico

Drug Trafficking, Organized Crime, and Violence in Mexico[7]

Mexico is a major producer and supplier to the U.S. market of heroin, methamphetamine, and marijuana and a major transit country for more than 95% of the cocaine sold in the United States.[8] Mexico is also a consumer of illicit drugs, particularly in northern states where criminal organizations have been paying their workers in product rather than in cash. Illicit drug use in Mexico increased from 2002 to 2008, and then remained relatively level from 2008 to 2011. According to the 2011 *National Drug Threat Assessment (NDTA)*, Mexican drug trafficking organizations and their affiliates "dominate [in] the supply and wholesale distribution of most illicit drugs in the United States" and are present more than 1,000 U.S. cities.[9]

The violence and brutality of the Mexican drug trafficking organizations escalated as they have battled for control of trafficking routes into the United States and local drug distribution networks in Mexico. U.S. and Mexican officials now often refer to the DTOs as transnational criminal organizations (TCOs) since they have branched out into other criminal activities, including human trafficking, kidnapping, armed robbery, and extortion. From 2007 to 2011, kidnapping and violent vehicular thefts increased at faster annual rates than homicides in Mexico.[10] The expanding techniques used by the traffickers, which have included the use of car bombs and grenades, have led some observers to liken certain DTOs' tactics to those of armed insurgencies.[11]

---

[7] For background, see CRS Report R41576, *Mexico's Drug Trafficking Organizations: Source and Scope of the Violence*, by June S. Beittel.

[8] U.S. Department of State, *International Narcotics Control Strategy Report (INCSR)*, March 2013, http://www.state.gov/j/inl/rls/nrcrpt/2013/vol1/204050 htm#Mexico. See also: U.S. Department of Justice (DOJ), Drug Enforcement Administration (DEA), *National Drug Threat Assessment (NDTA) Summary: 2013*, available at http://www.justice.gov/dea/resource-center/DIR-017-13%20NDTA%20Summary%20final.pdf.

[9] DOJ, National Drug Intelligence Center, *NDTA: 2011*, August 2011, http://www.justice.gov/ndic/pubs44/44849/44849p.pdf.

[10] From 2007 to 2011, the homicide rate per 100,000 people in Mexico increased by an annual average of 15.4%. During that same period, kidnappings increased at an average annual rate of 23.5% and armed vehicular robberies by 19.7%. Mexico Evalúa, *Indicadores de Víctimas Visibles y Invisiblesde Homicidio*, Mexico, D.F., November 2012, available at http://mexicoevalua.org/descargables/413537_IVVI-H.pdf.

[11] Robert J. Bunker and John P. Sullivan, "Cartel Evolution Revisited: Third Phase Cartel Potentials and Alternative Futures in Mexico," *Small Wars & Insurgencies*, vol. 21, no. 1 (March 2010).

The Felipe Calderón Administration (December 2006-November 2011) made combating drug trafficking and organized crime its top domestic priority. Government enforcement efforts, many of which were led by Mexican military forces, took down leaders from all of the major DTOs, either through arrests or deaths during operations to detain them. The pace of those takedowns accelerated beginning in late 2009, partly due to increased U.S.-Mexican intelligence-sharing. In 2009, the Mexican government identified the country's 37 most wanted criminals, and by October 2012, at least 25 of those alleged criminals had been captured or killed.[12] The Calderón government extradited record numbers of criminals to the United States, including 115 in 2012 (see **Figure 2**); however few, if any, were successfully prosecuted in Mexico.[13] At the same time, Mexico also experienced record levels of drug trafficking-related violence, partially in response to government efforts, as criminal organizations split and proliferated.

Drug trafficking-related violence in Mexico may have resulted in some 60,000 deaths over the course of Calderón's presidency; another 25,000 individuals reportedly went missing over that period, although not all due to criminal activity.[14] Several sources have reported that violence peaked in 2011, before falling in 2012. The violence took place largely in contested drug production and transit zones representing less than 10% of Mexican municipalities.[15]

Still, the regions of the country most affected by the violence have shifted over time to include large cities (such as Monterrey, Nuevo León) and tourist zones (Acapulco, Guerrero). There have been incidents of violence across the country, with the security situation in particular areas changing rapidly. For example, violence spiked dramatically in Ciudad Juárez, Chihuahua, in 2008 and remained at extremely high levels through mid-2011, before rapidly declining.

Upon taking office, President Peña Nieto's made violence reduction one of his top priorities. Organized crime-related violence continued to decline in 2013 as it had during the last year of the Calderón government, yet serious security challenges remain in many parts of Mexico. President Peña Nieto has said that organized-crime violence declined by 30% in 2013; experts have challenged the veracity of government figures.[16] Since the government is no longer publicly releasing information on trends in organized crime-related killings as opposed to all homicides, it is difficult to analyze the security situation with precision. According to Mexican government figures, all homicides fell by 16.5% as compared to 2012.[17] While violence has declined in some parts of northern Mexico (including Chihuahua, Baja California, and Nuevo León), it has spiked in the interior of the country and along the Pacific Coast, particularly in Michoacán and neighboring Guerrero. A 2014 State Department Travel Warning cited security concerns in parts of 19 of Mexico's 32 states.[18]

---

[12] "Mexico's Drug Lords: Kingpin Bowling," *The Economist*, October 20, 2012.

[13] William Booth, "Mexico's Crime Wave has Left About 25,000 Missing, Government Documents Show," *Washington Post*, November 29, 2012.

[14] TBI, February 2013; Booth op. cit. Mexico's Attorney General's Office is investigating how many of the disappearances may have been linked to organized crime or rogue government actors.

[15] TBI, February 2013.

[16] "Mexico Poised to Take-Off, Peña Nieto Tells Davos," *Latin News Daily*, January 24, 2014; Alejandro Hope, "Homicidios: Algo no Cuadra," *Animal Político*, March 25, 2014.

[17] Rafael Cabrera, "Menos Homicidios y más Secuestros Durante 2013: SNSP," *Animal Político*, January 24, 2014.

[18] U.S. Department of State, Bureau of Consular Affairs, *Travel Warning: Mexico*, January 4, 2014, available at http://travel.state.gov/content/passports/english/alertswarnings/mexico-travel-warning.html.

Figure I. Map of Mexico

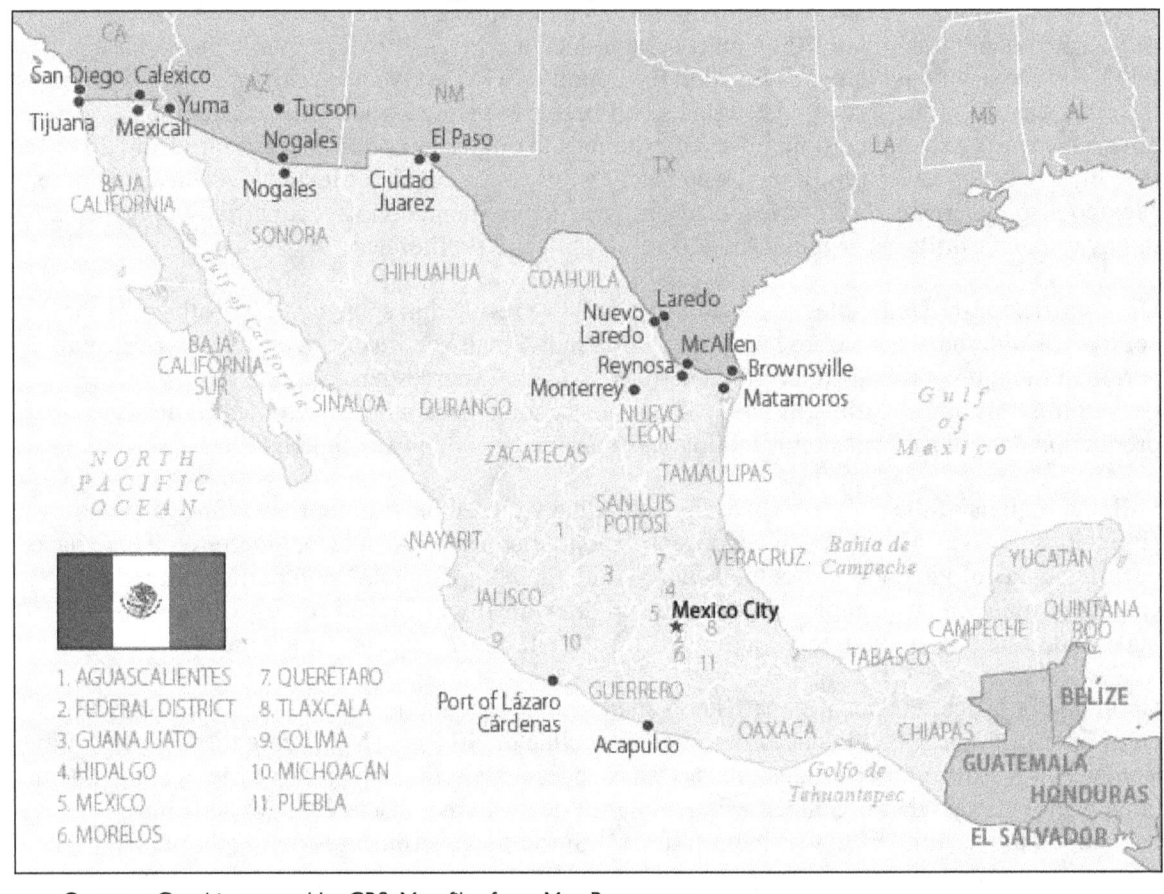

Source: Graphic created by CRS. Map files from Map Resources.

Drug Trafficking-Related Violence in the United States[19]

The prevalence of drug trafficking-related violence in northern Mexico between 2007 and 2011generated concern among U.S. policy makers that this violence might spill over into the United States. U.S. officials denied that the increase in drug trafficking-related violence in Mexico resulted in a significant spillover of violence into the United States, but acknowledged that the prospect was a real concern.[20] As drug trafficking-related violence has declined in Mexico, so too have concerns about the potential for spillover violence. Nonetheless, policy makers and experts have monitored the activities of drug trafficking organizations, both in Mexico and in the United States. Countering the movement of illegal drugs from Mexico into the U.S. market has remained a top U.S. drug control priority.[21]

---

[19] For background, see CRS Report R41075, *Southwest Border Violence: Issues in Identifying and Measuring Spillover Violence*, by Kristin Finklea.

[20] See for example, Department of Homeland Security, "Remarks by Secretary Napolitano on Border Security at the University of Texas at El Paso," press release, January 31, 2011.

[21] See, for example, the Office of National Drug Control Policy, *National Drug Control Strategy 2013*, April 2013.

Congress faces several policy questions related to potential or actual spillover violence. One question involves assessing whether the level of violence between the drug trafficking organizations in Mexico affects either the level or nature of drug trafficking-related violence in the United States. Of note, violent drug trafficking-related crimes have previously existed and continue to exist throughout the United States. However, data currently available on these crimes do not allow analysts to determine whether or how these existing levels of drug trafficking-related violence in the United States have been affected by the surge of violence in Mexico.

Development and Implementation of the Mérida Initiative

Evolution of U.S.-Mexican Counterdrug Cooperation

The United States began providing Mexico with equipment and training to eradicate marijuana and opium poppy fields in the 1970s, but bilateral cooperation declined dramatically after Enrique Camarena, a U.S. Drug Enforcement Administration (DEA) agent, was assassinated in Mexico in 1985. From the mid-1980s through the end of the 1990s, bilateral cooperation stalled due to U.S. mistrust of Mexican counterdrug officials and concerns about the Mexican government's tendency to accommodate drug leaders.[22] At the same time, the Mexican government was reluctant to accept large amounts of U.S. assistance due to its opposition to U.S. drug certification procedures[23] and to concerns about sovereignty. The Mexican government also expressed opposition to the DEA carrying out operations against drug trafficking organizations in Mexican territory without authorization. Mexican military officials proved particularly reticent to cooperate with the U.S. military due to concerns about past U.S. interventions in Mexico.[24]

U.S.-Mexican cooperation began to improve and U.S. assistance to Mexico increased after the two countries signed a Binational Drug Control Strategy in 1998. U.S. assistance to Mexico, which totaled $397 million from FY2000 to FY2006, supported programs aimed at interdicting cocaine; combating production and trafficking of marijuana, opium poppy, and methamphetamine; strengthening the rule of law; and countering money-laundering. In 2007, the Government Accountability Office (GAO) found that while U.S. programs had helped improve Mexico's counterdrug efforts, drug seizures in Mexico remained relatively low, and corruption continued to hinder bilateral efforts.[25]

---

[22] Under this system, arrests and eradication took place, but due to the effects of widespread corruption, the system was "characterized by a working relationship between Mexican authorities and drug lords" through the 1990s. Francisco E. González, "Mexico's Drug Wars Get Brutal," *Current History*, February 2009.

[23] Beginning in 1986, when the U.S. President was required to certify whether drug producing and drug transit countries were cooperating fully with the United States, Mexico usually was criticized for its efforts, which in turn led to increased Mexican government criticism of the U.S assessment. Reforms to the U.S. drug certification process enacted in September 2002 (P.L. 107-228) essentially eliminated the annual drug certification requirement, and instead required the President to designate and withhold assistance from countries that had "failed demonstrably" to make substantial counternarcotics efforts.

[24] Craig A. Deare, "U.S.-Mexico Defense Relations: An Incompatible Interface," *Strategic Forum*, Institute for National Strategic Studies, National Defense University, July 2009.

[25] U.S. Government Accountability Office (GAO), *U.S. Assistance Has Helped Mexican Counternarcotics Efforts, but the Flow of Illicit Narcotics into the United States Remains High*, 08215T, October 2007, available at (continued...)

Developing Cooperation through the Mérida Initiative

In October 2007, the United States and Mexico announced the Mérida Initiative, a package of U.S. assistance for Mexico and Central America that would begin in FY2008.[26] The Mérida Initiative was developed in response to the Calderón government's unprecedented request for increased U.S. support and involvement in helping Mexico combat drug trafficking and organized crime. As part of the Mérida Initiative's emphasis on *shared responsibility*, the Mexican government pledged to tackle crime and corruption and the U.S. government pledged to address drug demand and the illicit trafficking of firearms and bulk currency to Mexico.

The Mérida Initiative, as it was originally conceived, sought to (1) break the power and impunity of criminal organizations; (2) strengthen border, air, and maritime controls; (3) improve the capacity of justice systems in the region; and (4) curtail gang activity and diminish local drug demand. U.S. funds provided for the first goal far surpassed all other aid categories. The U.S. government also provided extensive intelligence-sharing and operational support for Mexican military and police personnel engaged in anti-crime efforts.

Acknowledging that Mexico cannot effectively confront organized crime with tactical victories alone, in March 2010, the Obama Administration and the Mexican government agreed to a new strategic framework for security cooperation under the Mérida Initiative.[27] Whereas U.S. assistance initially focused on training and equipping Mexican security forces for counternarcotic purposes, it has shifted toward addressing the weak government institutions and underlying societal problems that have allowed the drug trade to thrive in Mexico. The strategy focuses more on institution-building than on technology transfers and broadens the scope of bilateral efforts to include economic development and community-based social programs, areas where Mexico had not previously sought U.S. support. There is also increasing funding at the sub-national level for Mexican states and municipalities. The four pillars of the current strategy are:

1. Disrupting the operational capacity of organized criminal groups.

2. Institutionalizing reforms to sustain the rule of law and respect for human rights).

3. Creating a 21st century border.

4. Building strong and resilient communities.

U.S. and Mexican officials have described the Mérida Initiative as a "new paradigm" for bilateral security cooperation. As part of Mérida, the Calderón government put sovereignty concerns aside to allow extensive U.S. involvement in Mexico's domestic security efforts. The two governments increased cooperation through the establishment of a multi-level working group structure to design and implement bilateral security efforts that included annual cabinet-level meetings. A cabinet-level meeting did not occur during the first year of the Peña Nieto government.

(...continued)

http://www.gao.gov/new.items/d071018.pdf.

[26] In FY2008 and FY2009, the Mérida Initiative included U.S. assistance to Mexico and Central America. Beginning in FY2010, Congress separated Central America from the Mexico-focused Mérida Initiative by creating a separate Central American Regional Security Initiative (CARSI).

[27] U.S. Department of State, "Joint Statement of the Mérida Initiative High-Level Consultative Group on Bilateral Cooperation Against Transnational Organized Crime," March 29, 2010.

Nevertheless, Presidents Obama and Peña Nieto reaffirmed their commitments to the Mérida Initiative's four pillar strategy during President Obama's trip to Mexico in May 2013. In August 2013, the U.S. and Mexican governments then agreed to focus on "justice sector reform, efforts against money laundering, police and corrections professionalization at the federal and state level, border security both north and south, and piloting approaches to address root causes of violence."[28] For a detailed discussion of each of the pillars, including new areas of emphasis within those pillars, see "The Four Pillars of the Mérida Initiative" below.

Funding the Mérida Initiative

Congress, with the power of the purse, has played a major role in determining the level and composition of Mérida funding for Mexico. From FY2008 to FY2014, Congress appropriated more than $2.4 billion for Mexico under the Mérida Initiative (see **Table 1** for Mérida appropriations and **Table A-1** in **Appendix A** for overall U.S. assistance to Mexico). In the beginning, Congress included funding for Mexico in supplemental appropriations measures in an attempt to hasten the delivery of certain equipment. Congress has also earmarked funds in order to ensure that certain programs are prioritized, such as efforts to support institutional reform. From FY2012 onward, funds provided for pillar two have exceeded all other aid categories.

Congress has sought to influence human rights conditions and encourage efforts to combat abuses and impunity in Mexico by placing conditions on Mérida-related assistance. Congress directed that 15% of certain assistance provided to Mexican military and police forces would be subject to certain human rights conditions. The conditions included in the FY2014 Consolidated Appropriations Act (P.L. 113-76) are slightly different than in previous years (see "Human Rights Concerns and Conditions on Mérida Initiative Funding"). The explanatory statement[29] accompanying the Act also requires a report from the State Department within 60 days of the measure's enactment on progress made in meeting the human rights conditions included in FY2012 and FY2013 appropriations legislation (P.L. 112-74 and P.L. 113-6).[30]

Table 1. FY2008–FY2015 Mérida Funding for Mexico

($ in millions)

Account	FY2008	FY2009	FY2010	FY2011	FY2012	FY2013	FY2014 (est.)	Account Totals	FY2015 Request
ESF	20.0	15.0	15.0ᵃ	18.0	33.3	32.1	46.1	**179.5**	35.0
INCLE	263.5	406.0	365.0	117.0	248.5	195.1	148.1	**1,743.2**	80.0
FMF	116.5	299.0	5.3	8.0	N/Aᵇ	N/A	N/A	**428.8**	N/A
Total	**400.0**	**720.0**	**385.3**	**143.0**	**281.8**	**227.2**	**194.2**	**2,351.5**	**115.0**

---

[28] Testimony of John Feeley, Principal Deputy Assistant Secretary of State, Bureau of Western Hemisphere Affairs, before the U.S. Congress, House Committee on Homeland Security, *Taking Down the Cartels: Examining United States-Mexican Cooperation*, 113th Cong., 2nd sess., April 2, 2014.

[29] The statement is available here: http://rules.house.gov/bill/113/hr-3547-sa.

[30] The reporting requirement originally appeared in H.Rept. 113-185 accompanying the House Appropriations Committee's version of the FY2014 State-Foreign Operations appropriations bill, H.R. 2855.

Sources: U.S. Department of State, *Congressional Budget Justification for Foreign Operations FY2008-FY2015*.

Notes: ESF=Economic Support Fund; FMF=Foreign Military Financing; INCLE=International Narcotics Control and Law Enforcement.

a. $6 million was later reprogrammed for global climate change efforts by the State Department.

b. Beginning in FY2012, FMF assistance is not included as part of the Mérida Initiative.

Implementation

Over the past few years, Congress has maintained an interest in ensuring that Mérida-funded equipment and training is delivered efficiently.[31] After initial delays in 2009-2010, deliveries accelerated in 2011, a year in which the U.S. government provided Mexico more than $500 million worth of equipment, training, and technical assistance. As of November 2012, some $1.1 billion worth of assistance had been provided. That total included roughly $873.7 million in equipment (including 21 aircraft[32] and more than $100 million worth of non-intrusive inspection equipment) and $146.0 million worth of training.

As of February 2014, deliveries had inched upwards to $1.2 billion,[33] with roughly $50.7 million worth of training and equipment delivered in 2013. While $95 million or so in FY2012 INCLE funding is being withheld due to a congressional hold, significant Mérida funding appropriated has yet to be delivered. For most of 2013, delays in implementation occurred partially due to the fact that the Peña Nieto government was still honing its security strategy and determining the amount and type of U.S. assistance needed to support that strategy. The initial procedure the government adopted for processing all requests from Mexican ministries for Mérida support through the Interior Ministry also reportedly contributed to implementation delays.[34]

By November 2013, the State Department and Mexican Foreign Affairs and Interior Ministries had agreed to a new, more agile process for approving new Mérida Initiative projects. By mid-March 2014, the governments had agreed to some $309 million worth of new projects, including $95 million worth of projects in the states.[35]

U.S. assistance has increasingly focused on supporting efforts to strengthen institutions in Mexico through training and technical assistance. U.S. funds support training courses offered in new or refurbished training academies for customs personnel, corrections staff, canine teams, and police (federal, state, and local).[36] Some of that training is designed according to a "train the trainer"

---

[31] In H.Rept. 112-494, House appropriators maintained that Congress "continues to be concerned with the delivery of assistance to Mexico" and urge agencies to "use all appropriate means necessary to ensure the prompt delivery of equipment and training."

[32] Aerial equipment deliveries thus far have included four CASA 235 maritime surveillance aircraft, nine UH-60 Black Hawk helicopters, and eight Bell 412 helicopters. The only pending aircraft delivery is an Intelligence Surveillance, and Reconnaissance (ISR) Dornier 328-JET aircraft that has been contracted for the Federal Police.

[33] U.S. Embassy in Mexico City, "Fact Sheet: The Mérida Initiative - An Overview," February 2014, available at http://photos.state.gov/libraries/mexico/310329/feb2014/Merida-Initiative-Overview-2-14.pdf. Hereinafter, U.S. Embassy, February 2014.

[34] CRS interviews with Mexican analysts in Mexico City, November 19-20, 2013.

[35] Electronic correspondence with State Department official, March 31, 2014.

[36] Mérida assistance is also supporting Mexican institutions like the National Public Security System (SNSP), which sets police standards and provides grants to states and municipalities for police training, and the National Institute of Criminal Sciences (INACIPE), which provides training to judicial sector personnel.

model in which the academies train instructors who in turn are able to train their own personnel. As of May 2013, some 19,000 law enforcement officers (including 4,000 federal police investigators) had completed U.S. courses. Another 8,500 federal and 22,500 state justice sector personnel had received training on their roles in Mexico's new accusatorial justice system.[37] Despite these numbers, high turnover rates within Mexican criminal justice institutions, particularly since the transition from a PAN to a PRI government has limited the impact of some U.S. training programs.[38]

U.S. Efforts to Complement the Mérida Initiative

In the 2007 U.S.-Mexico joint statement announcing the Mérida Initiative, the U.S. government pledged to "intensify its efforts to address all aspects of drug trafficking (including demand-related portions) and continue to combat trafficking of weapons and bulk currency to Mexico."[39] Although not funded through the Mérida Initiative, the U.S. government has made efforts to address each of these issues (see **Appendix B** for how those efforts have advanced); some have been more successful than others. When debating future support for the Mérida Initiative, Congress may consider whether to simultaneously provide additional funding for these or other domestic activities that would enhance the United States' abilities to fulfill its pledges.

The Peña Nieto Administration's Security Strategy and the Mérida Initiative

PRI President Enrique Peña Nieto, a former governor of the state of Mexico, took office on December 1, 2012. Upon his inauguration, the centrist PRI, a nationalistic party that governed Mexico from 1929 to 2000, retook the presidency after 12 years of rule by the conservative National Action Party (PAN). The PRI also controls a plurality (but not a majority) in Mexico's Senate and Chamber of Deputies.

Upon his inauguration, President Peña Nieto announced a reformist agenda with specific proposals under five broad categories: (1) reducing violence; (2) combating poverty; (3) boosting economic growth; (4) reforming education; and (5) fostering social responsibility. The aim of those proposals is to bolster Mexico's competitiveness. Leaders from the PAN and leftist Party of the Democratic Revolution (PRD) signed on to President Peña Nieto's "Pact for Mexico," an agreement aimed at advancing the reform agenda. The Pact facilitated the package of several key constitutional reforms before falling apart in December 2013.[40]

On December 17, 2012, President Peña Nieto outlined a security policy that involves binding commitments from all levels of government. The six pillars of the strategy include (1) planning;

---

[37] U.S. Congress, House Committee on Foreign Affairs, Subcommittee on the Western Hemisphere, *U.S.-Mexico Security Cooperation: An Overview of the Merida Initiative 2008–Present*, 113th Cong., 1st sess., *CQ Congressional Transcripts*, May 23, 2013. Hereinafter *An Overview of the Merida Initiative*, May 2013.

[38] CRS interviews with Mexican analysts in Mexico City, May 6-8, 2013.

[39] U.S. Department of State and Government of Mexico, "Joint Statement on the Mérida Initiative: A New Paradigm for Security Cooperation," October 22, 2007.

[40] For general information on the Peña Nieto Administration, see CRS Report R42917, *Mexico: Background and U.S. Relations*, by Clare Ribando Seelke.

---

(2) prevention; (3) protection and respect of human rights; (4) coordination; (5) institutional transformation; and (6) monitoring and evaluation. Two priority proposals Peña Nieto sought to achieve in the security realm that have been accomplished included launching a national crime prevention plan and establishing a unified code of criminal procedures to cover judicial procedures for the federal government and the states. Still, the results of the overall strategy have been mixed. While organized crime-related homicides continued to trend downward in 2013 as they had since mid-2011, extortions and kidnappings surged. Key kingpins have been arrested, but armed civilian self-defense groups[41] have spread throughout Mexico and the government's has struggled to quell unrest in Michoacán (see below).

In order to better coordinate security efforts, President Peña Nieto secured approval from the Mexican Congress to place the Secretariat of Public Security (that included the Federal Police) and intelligence functions under the Interior Ministry. That ministry is now the focal point for security collaboration and intelligence-sharing with foreign governments, as well as with coordination with state and municipal authorities. The states have in turn been divided into five geographic regions and are being encouraged, but not required, to stand up unified state police commands to coordinate with federal forces.

In addition to strengthening the role of the Interior Ministry in security efforts, the Peña Nieto government envisions a revamped and modernized Attorney General's Office (PGR). Per reforms enacted in December 2013, the PGR will eventually be replaced by an independent Prosecutor General's Office. Peña Nieto's security strategy calls for accelerated implementation of the judicial reforms passed in 2008, a key priority of pillar two (institutional reform) of the Mérida Initiative. It also calls for a reduced usage of preventive detention and prison reform that includes rehabilitation and reinsertion. Some experts predicted that the implementation of judicial reform will hasten now that the Mexican Congress has approved a unified code of criminal procedure, which was promulgated in early March 2014, while others are not so certain.[42]

Peña Nieto's security strategy explicitly prioritizes human rights, citizen participation, and crime prevention; this could portend an increase in bilateral efforts under Mérida's pillar two (to protect human rights through institutional reform) and pillar four (to build resilient communities). Peña Nieto's strategy pledges to increase victims' assistance as per the Victim's Law enacted in January 2013, transfer cases of military abuses against civilians to civilian courts, and find missing persons while also preventing future disappearances. Human rights groups have been critical of government efforts to translate that rhetoric into reality, however.[43] The government launched a national prevention program focused on 57 high-crime communities with a $9 billion budget for 2013 that included socioeconomic, education, infrastructure, and drug treatment

---

[41] There is a tradition of rural self-defense groups or community police in more remote and often indigenous parts of Mexico because of a lack of access to police and other justice sector services in those communities. For example, in parts of Guerrero, the Regional Coordinator of Community Authorities (CRAC is the Spanish acronym), was formed by indigenous communities to establish a police force and a system of indigenous justice in the mid-1990s. The newer self-defense groups that spread rapidly in 2013 are more ad hoc and not necessarily tied to indigenous communities. They pursue criminal groups to other towns and cities; are self-appointed, sometimes gaining recruits who are former migrants returned from or deported from the United States; and many are heavily armed.

[42] For information on the pros and cons of the new code, see Viridiana Rios, "Justice in Mexico: "The Mexican Drug War's Most Important Change that Nobody Noticed," *Harvard Kennedy School Review,* March 26, 2014.

[43] Maureen Meyer and Clay Boggs, "One Year into Mexican President Enrique Peña Nieto's Administration: Little Progress has been Made on Security or Human Rights," press release, Washington Office on Latin America (WOLA), November 27, 2013.

---

programs. The program has been criticized, however, for lacking a rigorous methodology for selecting and evaluating the communities and interventions that it is funding.[44]

While U.S. and Mexican interests coalesced around security concerns along the border during the Calderón Administration, they have focused more on how to promote economic dynamism under pillar three of the Mérida Initiative (creating a 21st century border) since Peña Nieto took office. On September 20, 2013, Peña Nieto and Vice President Joseph Biden announced plans to enhance cooperation in border trade and security as part of the first annual U.S.-Mexico High-Level Economic Dialogue. Describing the U.S.-Mexican border as the "busiest in the world," President Peña Nieto stated that a top goal is to streamline trade and improve border crossing infrastructure so that the transit of both people and trade will become more efficient, faster, and safer.[45] Both governments are also discussing ways to expand pillar three efforts to Mexico's southern borders.

Some details of Peña Nieto's security strategy that will have implications for U.S.-Mexican cooperation have yet to be well defined. For example, the strategy envisions a continued role for the Mexican military in public security efforts through at least 2015; whether and how the role of the military will be different than under the Calderón government still needs to be clarified. The Federal Police is being reformed, rather than dismantled, but how the force will be reconfigured to focus on investigations and combating key crimes (such as kidnapping and extortion) remains to be seen. In addition to a reconfigured Federal Police, President Peña Nieto initially proposed to create a new militarized police entity with some 50,000 officers, the National Gendarmerie, whose forces were to be drawn from the military but placed under the control of the Interior Ministry. That proposal has since been scaled back to consist of a new unit within the Federal Police composed of some 5,000 officers that will begin operating in mid-2014.

In general, the Peña Nieto government's approach to security has been described as more low profile than that of former President Calderón, who publicized kingpin arrests and drug seizures and highlighted U.S.-Mexican joint operations. While that may describe his public relations approach, some analysts maintain that Peña Nieto has quietly maintained a security approach similar to that of Calderón, including a high level of cooperation with the United States.[46] Indeed, despite some restrictions placed on U.S. security agencies working in Mexico, U.S. intelligence reportedly helped Mexican marines successfully track and arrest Miguel Angel Treviño Morales ("Z-40"), the leader of Los Zetas, in July 2013.[47] The Mexican government arrested some 69 other top drug traffickers in 2013.[48] U.S. intelligence, wiretaps, and surveillance equipment, along with embedded Drug Enforcement Administration (DEA) agents and U.S. Marshals, recently helped the Mexican marines (SEMAR) track and capture Joaquín "El Chapo" Guzmán—the world's most wanted drug trafficker—without a shot being fired.[49]

---

[44] México Evalua, *Prevención del Delito en México: Dónde Quedó la Evidencia?*, January 2014, available at http://www.mexicoevalua.org/wp-content/uploads/2014/01/MEX-EVA_INDX-PREVDEL-LOW.pdf

[45] Maja Wallengren, "Biden, Mexico's Peña Nieto Inaugurate new Initiative to Enhance Trade, Cooperation," *International Trade Reporter*, September 24, 2013.

[46] Alfredo Méndez, "Peña Nieto Mantiene el Errático Plan de Seguridad de Calderón, Dicen Juristas, *La Jornada*, July 30, 2013.

[47] "No Shots Fired: Leader of Mexico's Zetas Cartel Captured in Precision Operation, with U.S. Help," *Associated Press*, July 16, 2013.

[48] Dudley Althaus, "Mexico 2013 Target List: Many Zetas, Little Impact," *In Sight Crime: Organized Crime in the Americas,* December 23, 2013.

[49] Damien Cave, "How a Kingpin Above the Law Fell, Incredibly, Without a Shot," *New York Times*, February 23, 2014; Joshua Partlow and Sari Horwitz, "U.S. and Mexican Authorities Detail Coordinated Effort to Capture Drug (continued...)

Michoacán: Confronting DTOs and Dealing with Self-Defense Groups[50]

What criminal groups operate in Michoacán?

During the Calderón Administration, the restive state of Michoacán demonstrated the limits of the strategy of deploying federal forces to go after kingpins. The situation in Michoacán has become even more complicated over the past year or so as civilian self-defense groups have armed themselves and sought to reclaim territory from the Knights Templar, the dominant criminal organization in the state. Like its hyper-violent predecessor, La Familia Michoacana (LFM), the Knights Templar claim to be an evangelical Christian, vigilante group fighting in order to "defend" the state from predatory drug cartels such as Los Zetas and other regional DTOs such as Jalisco New Generation Cartel (CJNG). The Knights Templar not only engage in drug trafficking, but have extracted rents from the Port of Lázaro Cárdenas, mining operations, farmers, and businesses throughout the state.

What are self-defense groups?

In Michoacán, the self-defense militias (sometimes described as vigilantes) consist of armed civilians, mainly from rural areas, who claim they are fighting because they have lost confidence in the Mexican government's willingness or capacity to combat the Knights Templar. Local businesses weary of extortion and violent crime have provided funding to the self-defense groups, but authorities fear they may be extending their ties to competing organized crime groups, such as the CJNG, to obtain weapons and resources. Several self-defense group leaders have been deported from the United States; one has recently been arrested for murder.[51] Another has refused a government order to disarm.[52]

What is the Federal Government's Policy in Michoacán?

During his first year in office, President Peña Nieto appeared to downplay security concerns that were brewing in certain states, including Michoacán. In November 2013, Peña Nieto sent Federal Police and military officials to take control of the Port of Lázaro Cárdenas from municipal police forces. By mid-January 2014, armed conflicts had broken out between the Knights Templar and self-defense groups. President Peña Nieto designated a special envoy for Michoacán, Alfredo Castillo, to coordinate federal efforts on behalf of the Ministry of the Interior with state and local authorities. Federal efforts included the deployment of 10,000 Federal Police and Army troops and the designation of some $3.4 billion in federal funding to be spent in the state. Rather than combating both the Knights Templar and forcibly disarming the self-defense groups simultaneously, the Peña Nieto government initially decided to absorb the civilian militias into "rural defense corps"[53] under the authority of the Army. Those who are registered, who reportedly numbered 900 individuals as of mid-March 2014,[54] are supposed to receive basic training, a uniform, and the ability to legally carry weapons as per an agreement reached with the government. They will not be paid.[55] The government has recently told the self-defense forces to disarm, but they have resisted. [56]

What has the strategy achieved thus far? The strategy has reportedly led to the arrest of more than 500 criminals and the killing of two of the leaders of the Knights Templar: Nazario Moreno, who had been erroneously reported as dead by the Calderón government, and Enrique Plancarte.[57]

(...continued)

Lord," *Washington Post*, February 23, 2014.

[50] June S. Beittel, Analyst in Latin American Affairs, contributed to this section.

[51] Alfredo Corchado, "Mexico's Ties to Vigilante Groups Unravel," *Dallas Morning News*, March 14, 2014.

[52] Richard Fausset and Cecilia Sanchez, "Mexican vigilante leader refuses government order to disarm," *Los Angeles Times*, April 7, 2014.

[53] According to Inigo Guevara, a Mexican defense analyst, the origins of the rural defense corps can be traced back to the 1860s when civilians were recruited to patrol roads near Mexico City. Although their mission has changed over time, they have usually been recruited on a temporary basis by military commanders in different zones to provide various forms of rural citizen protection.

[54] Nacha Cattan, "'El Americano's' 400 Gunmen Undermine Mexico Bid to Cut Violence," *Bloomberg*, March 28, 2014.

[55] "Mexico: Vigilantes to be Drafted into Security Organs," *Latin American Weekly Report*, January 30, 2014.

[56] Fausset, op. cit.

[57] Raúl Benitez Manaut, as cited in: Inter-American Dialogue, "Is Mexico Winning or Losing the Battle for the Rule of
(continued...)

> **Why is this strategy risky?** Many questions remain concerning this new policy. To what extent, if at all, is the Army vetting the self-defense groups for ties to crime groups? How is the Army ensuring that the groups do not get their weapons (at least new weapons) through illicit means? What type of training is being given to the newly-established "rural defense corps?" How is the Army exerting oversight over the actions of the rural defense corps? How will the groups react now that one of their leaders has been arrested? How long will the corps be needed in Michoacán and how will they be disbanded once they are no longer needed? How will the government prevent the rural defense groups from becoming permanent paramilitary forces?

The Four Pillars of the Mérida Initiative

Pillar One: Disrupting the Operational Capacity of Organized Crime

During the Calderón Administration, Mexico focused much of its efforts on dismantling the leadership of the major DTOs. U.S. assistance appropriated during the first phase of the Mérida Initiative (FY2008-FY2010) enabled the purchase of equipment to support the efforts of federal security forces engaged in anti-DTO efforts. That equipment included $590.5 million worth of aircraft and helicopters, as well as forensic equipment for the Federal Police and Attorney General's respective crime laboratories. As the DTOs continue to employ new weapons, new types of training and/or equipment may be needed to help security officials combat those new threats. U.S. surveillance equipment reportedly aided the Mexican marines in tracking and capturing both Miguel Angel Treviño Morales and Joaquín "El Chapo" Guzmán.[58] As the Peña Nieto government establishes a National Gendarmerie within the Interior Ministry and a Criminal Investigative Agency or AIC within the Attorney General's Office (PGR), increased assistance may be requested to assist those entities, as well as existing federal forces.

The Mexican government has increasingly been conceptualizing the DTOs as for-profit corporations. Consequently, its strategy, and U.S. efforts to support it, has begun to focus more attention on disrupting the criminal proceeds used to finance DTOs' operations, although much more could be done in that area.[59] In August 2010, the Mexican government imposed limits on the amount of U.S. dollars that individuals can exchange or deposit each month. In October 2012, the Mexican Congress approved an anti-money laundering law establishing a financial crimes unit within the PGR, subjecting industries vulnerable to money laundering to new reporting requirements, and creating new criminal offenses for money laundering. Future Mérida assistance could be used to provide additional equipment and technical assistance to units within the Finance Ministry and the PGR that are investigating money laundering cases.

As mentioned, the DTOs are increasingly evolving into poly-criminal organizations, perhaps as a result of drug interdiction efforts cutting into their profits. As a result, many have urged both

---

(...continued)

Law?" *Latin America Advisor*, March 20, 2014; "Mexican Marines Kill Templar Cartel's Leader," *Associated Press*, April 1, 2014.

[58] "No Shots Fired: Leader of Mexico's Zetas Cartel Captured in Precision Operation, with U.S. Help," *Associated Press*, July 16, 2013; Partlow and Horwitz, op. cit.

[59] Randal C. Archibold, "Vast Web Hides Mexican Drug Profits in Plain Sight, U.S. Authorities Say," *New York Times*, March 25, 2014.

---

governments to focus on combating other types of organized crime, such as kidnapping and alien smuggling. Some may therefore question whether the funding provided under the Mérida Initiative is being used to adequately address all forms of transnational organized crime.

Intelligence-sharing and cross-border law enforcement operations and investigations have been suggested as possible areas for increased cooperation. During the Calderón Administration, U.S. law enforcement and intelligence officials supported Mexican intelligence-gathering efforts in northern Mexico and U.S. drones gathered information that was shared with Mexican officials. A $13 million cross-border telecommunications system for sister cities along the U.S.-Mexico border that was funded by the Mérida Initiative is facilitating information-sharing among law enforcement operating in that region as well. Bilateral intelligence-sharing and law enforcement cooperation has continued under the Peña Nieto government, with U.S. assistance supporting the development of five regional intelligence centers.

A general question that may arise for policy makers as they review the Administration's budget requests for the Mérida Initiative is whether proposed funding would be used to expand existing bilateral partnerships (described in **Appendix B**), or to establish new partnerships. This may depend on the effectiveness of current partnerships, as well as whether new partnerships are needed to address emerging law enforcement challenges. As U.S. assistance increasingly flows to state-level law enforcement in Mexico, policy makers may consider if and to what extent those forces should participate in bilateral law enforcement partnerships. Finally, as Mexico receives U.S. equipment and training to secure its southern borders with Guatemala and Belize, the need for more regional partnerships with those countries might also arise.

Pillar Two: Institutionalizing Reforms to Sustain the Rule of Law and Respect for Human Rights in Mexico[60]

Reforming Mexico's corrupt and inefficient criminal justice system is widely regarded as a crucial for combating criminality, strengthening the rule of law, and better protecting citizen security and human rights in the country.[61] Due to concerns about the corruption and ineffectiveness of police and prosecutors, less than 13% of all crimes are reported in Mexico.[62] Even so, recent spikes in violence and criminality have overwhelmed Mexico's law enforcement and judicial institutions, with record numbers of arrests rarely resulting in successful convictions. Increasing cases of human rights abuses committed by authorities at all levels, as well as Mexico's inability to investigate and punish those accused of abuses, are also pressing concerns.

Federal police reform got underway during the Calderón Administration, although recent cases of police misconduct have highlighted lingering concerns about federal forces. A major challenge has been expanding police reform efforts to the state and municipal level, possibly through state-level unified police commands. Mérida funding has been used to extend U.S.-funded federal police training efforts to police from all 32 states through a National Police Training Program.

---

[60] For more information on this pillar, see CRS Report R43001, *Supporting Criminal Justice System Reform in Mexico: The U.S. Role*, by Clare Ribando Seelke.

[61] David Shirk, *The Drug War in Mexico: Confronting a Shared Threat*, Council on Foreign Relations, March 2011, available at http://www.cfr.org/mexico/drug-war-mexico/p24262.

[62] Gobierto Federal de Mexico, Instituto Nacional de Estadísticas y Geografía (INEGI), *2012 Encuesta Nacional de Victimización y Percepción sobre Seguridad Pública.*

With impunity rates hovering around 82% for homicide and even higher for other crimes,[63] experts maintain that it is crucial for Mexico to implement the judicial reforms passed in the summer of 2008 and to focus on fighting corruption at all levels of government. In order for Mexico to transition its criminal justice system to an accusatorial system with oral trials by 2016, many have argued that U.S.-funded judicial training programs need to be expanded. And, while U.S. assistance has helped federal prisons expand and improve, thousands of federal prisoners are still being housed in state prisons that are overcrowded and often extremely insecure.[64]

Reforming the Police

Police corruption has presented additional challenges to the campaign against DTOs in Mexico. While corruption has most often plagued municipal and state police forces, in June 2012 corrupt Federal Police officers involved in running a drug smuggling ring out of the Mexico City airport killed three of their colleagues. Corrupt officials have also been dismissed from the PGR's organized crime unit, as well as its police force.

The Calderón Administration took steps to reform Mexico's police forces by dramatically increasing police budgets, raising selection standards, and enhancing police training and equipment at the federal level. It also created a national database through which police at all levels can share information and intelligence, and accelerated implementation of a national police registry. President Calderón initially proposed the creation of one unified federal police force under the Secretariat for Public Security (SSP), but two laws passed in 2009 created a Federal Police (FP) force under the SSP and a Federal Ministerial Police (PFM) force under the PGR, both with some investigative functions.[65] It took the Mexican government another year to issue regulations delineating the roles and responsibilities of these two police entities. It remains to be seen how the Peña Nieto government's placement of the SSP under the authority of the Interior Ministry, its creation of a new National Gendarmerie, and its decision to put the PFM within the PGR's investigative agency will affect bilateral efforts. U.S. officials have offered to help Mexico develop national policing standards.[66]

Whereas initiatives to recruit, vet, train, and equip the FP under the SSP rapidly advanced (with support from the Mérida Initiative), efforts to build the PGR's police force have lagged behind. According to the State Department, Mérida funding supported specialized training courses to improve federal police investigations, intelligence collection and analysis, and anti-money laundering capacity, as well as the construction of regional command and control centers.[67] The

---

[63] In other words, about 82% of perpetrators have not been brought to justice. Guillermo Zepeda, *Seguridad y Justicia Penal en los Estados: 25 Indicadores de Nuestra Debilidad Institucional*, Mexico Evalúa, March 2012.

[64] Federal prison reform in Mexico began in 2008. U.S. funding supported the refurbishment of a federal penitentiary academy in Veracruz and the accreditation of eight of México's federal facilities and five state prisons in Chihuahua by the American Correctional Association (ACA). U.S. training has heretofore been provided at the federal academy and the academy in Chihuahua, as well as in courses offered in Colorado and New Mexico. It is gradually being expanded in 2014 into Baja California, Baja California Sur, Coahuila, the Federal District, Nuevo León, the state of Mexico, Tamaulipas, and Veracruz, with a focus on bringing those academies up to ACA accreditation standards. Email from State Department official in Mexico City, March 28, 2014.

[65] Daniel Sabet, *Police Reform in Mexico: Advances and Persistent Obstacles*, Woodrow Wilson Center's Mexico Institute, Working Paper Series on U.S.-Mexico Security Cooperation, May 2010, available at http://www.wilsoncenter.org/topics/pubs/Sabet.pdf.

[66] *An Overview of the Mérida Initiative*, May 2013.

[67] U.S. Department of State, *FY2010 Mérida Initiative Spending Plan for Mexico*, June 10, 2010.

Calderón government also sought U.S. technical assistance in developing in-service evaluations and internal investigative units to prevent and punish police corruption and human rights abuses, although experts maintain that much more could be done in that area. Mérida assistance has also supported the PFM, although not to the same degree. U.S. training is now being provided to the PFM under the PGR's newly-established criminal investigative agency or AIC;[68] it could potentially support new FP units dedicated to combating kidnapping and extortion.

Thus far, state and local police reform has lagged behind federal police reform efforts. A public security law codified in January 2009 established vetting and certification procedures for state and local police to be overseen by the National Public Security System (SNSP). Federal subsidies have been provided to state and municipal units whose officers meet certain standards. Nevertheless, as of November 2012, the head of the SNSP at that time reported that only six states had complied with the 2009 law's requirement that all state and municipal police officers be vetted by January 2013; that deadline since has been extended. Still, concerns have been raised about the tests' reliability. And, even in states where vetting requirements have been met, a significant percentage of officers who failed the tests have remained on the job.[69]

The establishment of unified state police commands that could potentially absorb municipal police forces has been debated in Mexico for years.[70] The Mexican Congress failed to pass a constitutional reform proposal put forth by the Calderón government to establish unified state police commands. Nevertheless, President Peña Nieto is helping states move in that direction. In the meantime, some states have moved forward with plans to do away with municipal forces.

The outcome of the aforementioned reform efforts could have implications for U.S. initiatives to expand Mérida assistance to state and municipal police forces, particularly as the Mexican government determines how to organize and channel that assistance. Mérida funding has supported state-level academies in Chihuahua, Nuevo León, Puebla, and Sonora. The U.S. and Mexican governments have agreed to expand U.S. support to state academies in Chiapas, the Federal District, the state of Mexico, Michoacán, and Veracruz.[71] In 2013, Mérida funds supported training courses for 2,000 state and local police in officer safety, securing crime scene preservation, investigation techniques, and intelligence-gathering.

In order to complement these efforts, analysts have maintained that it is important to provide assistance to civil society and human rights-related non-governmental organizations (NGOs) in Mexico in order to strengthen their ability to monitor police conduct and provide input on policing policies. Some maintain that citizen participation councils, combined with internal control mechanisms and stringent punishments for police misconduct, can have a positive impact on police performance and police-community relations. Others have mentioned the importance of establishing citizen observatories to develop reliable indicators to track police and criminal justice system performance, as has been done in some states.

---

[68] "La PGR crea la nueva Agencia de Investigación Criminal," *CNN México*, September 25, 2013.

[69] "Mexico: the Tough Task of Cleaning up the Police," *Latin American Weekly Report*, November 15, 2012.

[70] Proponents of the reform maintain that it would improve coordination with the federal government and bring efficiency, standardization, and better trained and equipped police to municipalities. Skeptics argue that police corruption has been a major problem at all levels of the Mexican policing system and argue that there is a role for municipal police who are trained to deal with local issues.

[71] Electronic correspondence with State Department official, March 28, 2014.

Reforming the Judicial and Penal Systems

The Mexican judicial system has been widely criticized for being opaque, inefficient, and corrupt. It is plagued by long case backlogs, a high pre-trial detention rate, and an inability to secure convictions. The vast majority of drug trafficking-related arrests that have occurred over the last six years have not resulted in successful prosecutions. The PGR has also been unable to secure charges in many high-profile cases involving the arrests of politicians accused of collaborating with organized crime, such as Gregorio Sanchez, the former mayor of Cancun.[72]

Mexican prisons, particularly at the state level, are also in need of significant reforms. Increasing arrests have caused prison population to expand significantly, as has the use of preventive detention. Those suspected of involvement in organized crime can be held by the authorities for 40 days without access to legal counsel, with a possible extension of another 40 days, a practice known as "*arraigo*" (pre-charge detention) that has led to serious abuses by authorities.[73] Mexico's Attorney General has spoken out against the excessive use of *arraigo*, but the government continues to say it is necessary to facilitate some types of investigations.[74] Many inmates (perhaps 40%) are awaiting trials, as opposed to serving sentences. As of July 2013, prisons were at 22% over-capacity.[75] Prison breaks are common in state facilities, many of which are controlled by crime groups.

In June 2008, then-President Calderón signed a judicial reform decree after securing the approval of Congress and Mexico's states for an amendment to Mexico's Constitution. Under the reform, Mexico has until 2016 to replace its trial procedures at the federal and state level, moving from a closed-door process based on written arguments to a public trial system with oral arguments and the presumption of innocence until proven guilty. In addition to oral trials, judicial systems are expected to adopt additional means of alternative dispute resolution, which should help make it more flexible and efficient, thereby relieving some of the pressure on the country's prison system. To implement the reforms, Mexico will need to implement the unified code of criminal procedure at the federal and state level, build new courtrooms, retrain current legal professionals, update law school curricula, and improve forensic technology—a difficult and expensive undertaking.

Six years into the reform process, progress had, until recently, stalled at the federal level. From the beginning, analysts had predicted that progress in advancing judicial reform was "likely to be very slow as capacity constraints and entrenched interests in the judicial system (including judges) delay any changes."[76] Still, the Calderón government devoted more attention towards

---

[72] Patrick Corcoran, "Mayor Goes Free, Mexico Fails Again to Prosecute 'Corrupt' Politicians," *In Sight Organized Crime in the Americas,* July 21, 2011.

[73] This practice first came into existence in the 1980s, and was formally incorporated into the Mexican Constitution through a constitutional amendment passed in 2008 as a legal instrument to fight organized crime. Its use has been criticized by several United Nations bodies, the Inter-American Commission for Human Rights of the Organization of American States, and international and Mexican human rights organizations. For more, see Janice Deaton, *Arraigo and Legal Reform in Mexico*, University of San Diego, June 2010.

[74] U.S. Department of State, *Country Report on Human Rights Practices for 2013: Mexico*, February 2014, http://www.state.gov/j/drl/rls/hrrpt/humanrightsreport/#wrapper. Hereinafter: *Country Report: Mexico*, February 2014. Tanya Montalvo, "Para Proteger el Éxito de una Investigación": así Defiende México al Arraigo," *Animal Político*, March 31, 2014.

[75] *Country Report: Mexico*, February 2014.

[76] "Mexico Risk: Legal and Regulatory Risk," *Economist Intelligence Unit-Risk Briefing*, January 8, 2010.

modernizing the police than strengthening the justice system.[77] In addition, some of the tough measures for handling organized crime cases it included in the 2008 judicial reforms appear to run counter to the spirit of the reforms, which include protections for the rights of the accused.[78] Former President Calderón proposed a new federal criminal procedure code (CPC)—a key element needed to guide reform efforts—in September 2011, but it was not enacted.

President Peña Nieto has repeatedly pledged to advance judicial reform and overhaul the PGR. The Mexican Congress approved a unified code of criminal procedure to cover the entire judicial system in February 2014; it was promulgated in early March, 2014. Experts have guardedly praised that development, but remain unsure of how it will affect reform efforts, particularly in states that had already adopted new codes.[79]

In contrast to this lack of progress at the federal level, the reform has moved forward in many Mexican states. As of August 2013, 26 of Mexico's 32 states had enacted legislation to begin the transition to an oral and adversarial justice system and 16 states had begun operating at least partially under the new system.[80] Reform states have seen positive initial results as compared to non-reform states: faster case resolution times, less pre-trial detention, and tougher sentences for cases that go to trial.[81] Still, daunting challenges remain, including the need to improve the investigative capacity of police and prosecutors, counter-reform efforts, and opposition from judges and other key justice sector operators.

The U.S. Agency for International Development (USAID) is concentrating most of its work in support of judicial reform at the state level. USAID had been supporting code reform, judicial exchanges, alternative dispute resolution, and Citizen's Participation Councils, as well as training justice sector operators in five Mexican states since 2004. USAID expanded its rule of law efforts with roughly $104 million in FY2008-FY2012 Mérida assistance, a significant portion of which is supporting comprehensive judicial reform programs in 7 of Mexico's 32 states. USAID plans to expand its programs into at least 13 other states, with assistance to each state tailored to the stage that it is at in the reform process. USAID and the State Deparment also provide technical assistance to federal entity within the Interior Ministry that is charged with coordinating the federal and state level reform efforts, the Technical Secretariat for Criminal Justice Sector Reform Implementation (SETEC).

The Department of Justice (DOJ) has supported judicial reform at the federal level, including providing technical assistance to the Congress during the drafting and adoption of a unified CPC. DOJ has administered some $46 million in State Department funding. In 2012, DOJ worked with the PGR to design and implement a national training program known as *Project Diamante* through which 7,700 prosecutors, investigators, and forensic experts were trained to work as a team rather than in isolation (as was customary). DOJ also implemented a training program in

---

[77] Andrew Selee and Eric L. Olson, *Steady Advances, Slow Results: U.S.-Mexico Security Cooperation After Two Years of the Obama Administration*, Woodrow Wilson Center's Mexico Institute, April 2011.

[78] For a discussion of these concerns and the reform process in general, see David Shirk, "Criminal Justice Reform in Mexico: An Overview," *Mexican Law Review*, vol. 2, no. 3 (January-June 2011).

[79] Ríos, op. cit.

[80] *Country Report: Mexico*, February 2014.

[81] USAID, Justice Studies Center of the Americas, and Coordination Council for the Implementation of the Criminal Justice System and its Technical Secretariat (SETEC); *Monitoring the Implementation of the Criminal Justice Reform in Chihuahua, the State of Mexico, Morelos, Oaxaca, and Zacatecas: 2007-2011*, November 2012.

Puerto Rico for Mexican federal judges. Building on Project Diamante, the PGR is using Diamante instructors and the Diamante training management and evaluation team to begin a pilot program to transition its personnel and operations to the accusatorial system in Durango and Puebla. DOJ and PGR have a working group that will monitor the progress of the transition, evaluate the challenges faced, and disseminate lessons learns to PGR staff in other states.

Congress has expressed support for the continued provision of U.S. assistance for judicial reform efforts in Mexico in appropriations legislation, hearings, and committee reports. Congressional funding and oversight of judicial reform programs in Mexico is likely to continue for many years. Over time, Congress may consider how best to divide funding between the federal and state levels; how to sequence and coordinate support to key elements within the rule of law spectrum (police, prosecutors, courts); and how the efficacy of U.S. programs is being measured.

Pillar Three: Creating a "21st Century Border"

Policy makers have questioned not only what it means to have a 21st century border, but specifically how this will enhance law enforcement's abilities to combat the drug trafficking organizations and reduce the related violence. In an increasingly globalized world, the notion of a border is necessarily more complex than a physical line between two sovereign nations. Consequently, the proposed 21st century border is based on (1) enhancing public safety via increased information sharing, screenings, and prosecutions; (2) securing the cross-border flow of goods and people; (3) expediting legitimate commerce and travel through investments in personnel, technology, and infrastructure; (4) engaging border communities in cross-border trade; and (5) setting bilateral policies for collaborative border management.[82]

On May 19, 2010, the United States and Mexico declared their intent to collaborate on enhancing the U.S.-Mexican border.[83] A Twenty-First Century Border Bilateral Executive Steering Committee (ESC) met in December 2010, December 2011, and April 2013 to develop bi-national action plans for the subsequent year.[84] The plans are focused on setting measurable goals within broad objectives: coordinating infrastructure development, expanding trusted traveler and shipment programs, establishing pilot projects for cargo pre-clearance, improving cross-border commerce and ties, and bolstering information sharing among law enforcement agencies.

Both the United States and Mexico spend significant funds—outside of Mérida—related to border security. Because border policies and practices have been different along the U.S. side of the Southwest border and the Mexican side, each country's goals in further developing the border may necessarily differ as well. A related issue is whether funds appropriated under the revised Mérida Initiative should be divided equally or equitably between border initiatives on the U.S. and Mexican sides of the border.

---

[82] U.S. Department of State, "United States - Mexico Partnership: A New Border Vision," press release, March 23, 2010. See also U.S. Department of State, "United States - Mexico Partnership: Managing our 21st Century Border," Fact Sheet, April 30, 2013.

[83] The White House, "Declaration by The Government Of The United States Of America and The Government Of The United Mexican States Concerning Twenty-First Century Border Management," press release, May 19, 2010. As mentioned, U.S. - Mexican security cooperation along the border did not begin with the Mérida Initiative. This ESC is one of the most recent developments in the bilateral cooperation.

[84] More information on the 2013 Action Plan is available at https://www.dhs.gov/sites/default/files/publications/21cb-2013-action-plan.pdf.

While policy makers may generally question what constitutes a "21st century border," they may more specifically question which aspects of this border will be mutually beneficial to both U.S. and Mexican efforts to combat the DTOs. Although a key goal of the Mérida Initiative is to combat the DTOs and their criminal activities, the U.S. border strategy does not discriminate between combating drug trafficking-related illicit activities and other illegal behaviors along the border. The current U.S. border strategy strives to secure and manage the U.S. border through obtaining effective control of the borders, safeguarding lawful trade and travel, and identifying and disrupting transnational criminal organizations.[85] As such, it remains to be seen whether enhancements to the border will specifically support the Mérida Initiative's goal of combating the DTOs or whether the funds put toward border development will result in a general strengthening of the security of the border—and, as a byproduct, aid in disrupting drug trafficking activities.

Northbound and Southbound Inspections[86]

One element of concern regarding enhanced bilateral border security efforts is that of southbound inspections of people, goods, vehicles, and cargo. In particular, both countries have acknowledged a shared responsibility in fueling and combating the illicit drug trade. Policy makers may question who is responsible for performing northbound and southbound inspections in order to prevent illegal drugs from leaving Mexico and entering the United States and to prevent dangerous weapons and the monetary proceeds of drug sales from leaving the United States and entering Mexico. Further, if this is a joint responsibility, it is unclear how U.S. and Mexican border officials will divide the responsibility of inspections to maximize the possibility of stopping the illegal flow of goods while simultaneously minimizing the burden on the legitimate flow of goods and preventing the duplication of efforts.

In addition to its inbound/northbound inspections, the United States has undertaken steps to enhance its outbound/southbound screening procedures. Currently, DHS is screening 100% of southbound rail shipments for illegal weapons, cash, and drugs. Also, CBP scans license plates along the Southwest border with the use of automated license plate readers (LPRs). Further, CBP employs Non-Intrusive Inspection (NII) systems—both large-scale and mobile—to aid in inspection and processing of travelers and shipments. As of May 2013, CBP had "309 large-scale NII systems deployed to and in between U.S. ports of entry."[87]

Historically, Mexican Customs had not served the role of performing southbound (or inbound) inspections. As part of the revised Mérida Initiative, CBP has helped to establish a Mexican Customs training academy to support professionalization and promote the Mexican Customs' new role of performing inbound inspections. Additionally, CBP is assisting Mexican Customs in developing investigator training programs and the State Department has provided over 300 canines to assist with the inspections.[88]

---

[85] For a fuller discussion of U.S. border security policies, see CRS Report R42138, *Border Security: Immigration Enforcement Between Ports of Entry*, by Lisa Seghetti. CRS was unable to locate an official Mexican border strategy for comparison with the U.S. border strategy.

[86] There is a dearth of open-source data that currently measures the extent of inbound and outbound inspections performed by both the United States and Mexico along the Southwest border. Rather, existing data tend to address seizures of drugs, guns, and money as well as apprehensions of suspects. Therefore, this section addresses current U.S. and additional initiatives to bolster cross-border inspections.

[87] U.S. Customs and Border Protection, *Non-Intrusive Inspection (NII) Technology*, Fact Sheet, May 2013.

[88] Embassy of Mexico, *Fact Sheet: The Mérida Initiative—An Overview*, February 2014.

Preventing Border Enforcement Corruption

Another point that policy makers may question regarding the strengthening of the Southwest border is how to prevent the corruption of U.S. and Mexican border officials who are charged with securing the border. On March 11, 2010, the Senate Committee on Homeland Security and Governmental Affairs, Subcommittee on State, Local, and Private Sector Preparedness and Integration held a hearing on the corruption of U.S. border officials by Mexican DTOs. According to testimony from the hearing, in FY2009, the DHS Inspector General opened 839 investigations of DHS employees. Of the 839 investigations, 576 were of CBP employees, 164 were of ICE employees, 64 were of Citizen and Immigration Services (CIS) employees, and 35 were of Transportation Security Administration (TSA) employees.[89] It is unknown, however, how many of these cases involve alleged corruption by Mexican DTOs or how many involve suspected corruption of DHS employees working along the Southwest border. Data from a 2013 GAO report can provide a snapshot of the proportion of corruption involving Southwest border officials:

> From fiscal years 2005 through 2012, a total of 144 [CBP] employees were arrested or indicted for corruption-related activities, including the smuggling of aliens or drugs... About 65 percent (93 of 144 arrests) were employees stationed along the southwest border.[90]

To date, the Administration's proposal for a 21st century border has not directly addressed this issue of corruption. Congress may consider whether preventing, detecting, and prosecuting public corruption of border enforcement personnel should be a component of the border initiatives funded by the Mérida Initiative. If the corruption is as pervasive as officials say,[91] resources provided for new technologies and initiatives along the border may be diminished or negated by corrupt border personnel. For instance, at the end of 2009, CBP was able to polygraph between 10% and 15% of applicants applying for border patrol positions, and of those who were polygraphed, about 60% were found unsuitable for service.[92] If this pattern holds true and 85%-90% of current new hires were not subjected to a polygraph, anywhere between 51% and 54% of all CBP new-hires may not be found suitable for service. Further, between October 1, 2004, and March 11, 2010, 103 CBP officers were arrested or indicted on "mission-critical corruption charges including drug smuggling, alien smuggling, money laundering and conspiracy."[93] Further, in FY2011 and FY2012, sixteen CBP employees were arrested or indicted for "mission-

---

[89] See testimony by Thomas M. Frost, Assistant Inspector General for Investigations, U.S. Department of Homeland Security before the U.S. Congress, Senate Committee on Homeland Security and Governmental Affairs, Ad Hoc Subcommittee on State, Local, and Private Sector Preparedness and Integration, *New Border War: Corruption of U.S. Officials by Drug Cartels*, 111th Cong., 1st sess., March 11, 2010.

[90] U.S. Government Accountability Office, *Border Security: Additional Actions Needed to Strengthen CBP Efforts to Mitigate Risk of Employee Corruption and Misconduct*, GAO-13-59, December 2012, p. 9.

[91] See testimony by Kevin L. Perkins, Assistant Director, Criminal Investigative Division, Federal Bureau of Investigation before the U.S. Congress, Senate Committee on Homeland Security and Governmental Affairs, Ad Hoc Subcommittee on State, Local, and Private Sector Preparedness and Integration, *New Border War: Corruption of U.S. Officials by Drug Cartels*, 111th Cong., 1st sess., March 11, 2010.

[92] See testimony by James F. Tomsheck, Assistant Commissioner, Office of Internal Affairs, Customs and Border Protection before the U.S. Congress, Senate Committee on Homeland Security and Governmental Affairs, Ad Hoc Subcommittee on State, Local, and Private Sector Preparedness and Integration, *New Border War: Corruption of U.S. Officials by Drug Cartels*, 111th Cong., 1st sess., March 11, 2010.

[93] Ibid.

compromising corruption."[94] Congress may decide to increase funding—as part of or separately from Mérida funding—for the vetting of new and current border enforcement personnel.

Mexico's Southern Borders

Policy makers may also seek to examine a relatively new element under pillar three of the Mérida Initiative that involves U.S. support for securing Mexico's porous and insecure southern borders with Guatemala and Belize. The Mexican government has designed a southern border security plan that involves three security cordons that stretch more than 100 miles north of the Mexico-Guatemala and Mexico-Belize borders.[95] The State Department has provided $6.6 million of mobile Non-Intrusive Inspection Equipment (NIIE) and approximately $3.5 million in mobile kiosks, operated by Mexico's National Migration Institute, that capture the biometric and biographic data of individuals living and transiting southern Mexico. The U.S. Department of Defense (DOD) has also provided training to troops patrolling the border, communications equipment, and support for the development of Mexico's air mobility and surveillance capabilities (see "Role of the U.S. Department Of Defense in Mexico").

Pillar Four: Building Strong and Resilient Communities

This pillar is a new focus for U.S.-Mexican cooperation, the overall goals of which are to address the underlying causes of crime and violence, promote security and social development, and build communities that can withstand the pressures of crime and violence. Pillar four is unique in that it has involved Mexican and U.S. federal officials working together to design and implement community-based programs in high-crime areas in municipalities near the U.S.-Mexico border. Pillar four seeks to empower local leaders, civil society representatives, and private sector actors to lead crime prevention efforts in their communities.

In January 2010, in response to a violent massacre of 15 youth with no apparent connection to organized crime in Ciudad Juárez, Chihuahua, the Mexican government began to prioritize crime prevention and community engagement. Responding to criticisms of its military-led strategy for the city, federal officials worked with local authorities and civic leaders to establish six task forces to plan and oversee a strategy for reducing criminality, tackling social problems, and improving citizen-government relations. The strategy, aptly titled "Todos Somos Juarez" ("We Are All Juárez"), was launched in February 2010 and involved close to $400 million in federal investments in the city.[96] While federal officials began by amplifying access to existing social programs and building infrastructure projects throughout the city, they later sought to respond to local demands to concentrate efforts in certain "safe zones." At the same time, control over public security efforts in the city shifted from the Mexican military to the Federal Police, and finally to municipal authorities.[97]

---

[94] U.S. Government Accountability Office, *Border Security: Additional Actions Needed to Strengthen CBP Efforts to Mitigate Risk of Employee Corruption and Misconduct*, GAO-13-59, December 2012, p. 10.

[95] "Obama Administration Considers Plan to Bolster Mexico's Southern Border," *Washington Free Beacon*, August 22, 2013.

[96] Adam Thompson, "Troubled Juárez starts to breathe again," *Financial Times*, October 11, 2012.

[97] Each of these forces has committed human rights violations and exhibited corruption. Despite concerns about his aggressive tactics, the current municipal police chief in Ciudad Juárez has won praise by some for reducing crime rates. William Booth, "In Mexico's Murder City, the War Appears Over, *Washington Post*, August 20, 2012; Damien Cave, (continued...)

Prior to the endorsement of a formal pillar four strategy, the U.S. government's pillar four efforts in Ciudad Juárez involved the expansion of existing initiatives, such as school-based "culture of lawfulness"[98] programs and drug demand reduction and treatment services.[99] Culture of Lawfulness (CoL) programs aim to combine "top-down" and "bottom-up" approaches to educate all sectors of society on the importance of upholding the rule of law. U.S. support also included new programs, such as support for an anonymous tip line for the police. USAID supported a crime and violence mapping project[100] that enabled Ciudad Juarez's municipal government to identify hot spots and respond with tailored prevention measures as well as a program to provide safe spaces, activities, and job training programs for youth at-risk of recruitment to organized crime. USAID also provided $1 million in grants to local organizations working in the areas of social cohesion in Ciudad Juarez, with activities focused specifically on education, mental health, and at-risk youth, among others.

It may never be determined what role the aforementioned efforts played in the significant reductions in violence that has occurred in Ciudad Juaréz since 2011.[101] Nevertheless, lessons can be gleaned from this example of Mexican and U.S. involvement in municipal crime prevention. Analysts have praised the sustained, high-level support Ciudad Juárez received from the Mexican and U.S. governments; community ownership of the effort; and coordination that occurred between various levels of the Mexican government. The work of the security task force (*Mesa de Seguridad*) proved crucial for developing trust between citizens and authorities, communication among authorities, and citizen oversight of government efforts.[102] The strategy was not well-targeted, however, and monitoring and evaluation of its effectiveness has been relatively weak.[103]

In April 2011, the U.S. and Mexican governments formally approved a bi-national pillar four strategy.[104] The strategy focuses on three objectives: (1) strengthening federal civic planning capacity to prevent and reduce crime; (2) bolstering the capacity of state and local governments to implement crime prevention and reduction activities; and (3) increasing engagement with at-risk

---

(...continued)

"A Crime Fighter Draws Plaudits, and Scrutiny," *New York Times*, December 23, 2011.

[98] Key sectors that CoL programs seek to involve include law enforcement, security forces, and other public officials; the media; schools; and religious and cultural institutions. The U.S. government is supporting school-based "culture of lawfulness" programs, as well as "culture of lawfulness" courses that are being taught to federal and state police.

[99] U.S.-funded demand reduction programs are helping to create a network to connect Mexico's 334 prevention and treatment centers, to develop curricula for drug counselors at the centers, and to help certify Mexican drug counselors.

[100] The project has gathered available data on where violence is occurring in the city. See http://www.observatoriodejuarez.org/.

[101] While many analysts credit the decline in violence to the end of a turf war between the Sinaloa and Juárez DTOs, federal and local officials have variously taken credit for the reduction. See, for example, "Looking Back on the Calderón Years," *The Economist*, November 22, 2012.

[102] Lucy Conger, "The Private Sector and Public Security: The Cases of Ciudad Juarez and Monterrey," *Building Resilient Communities: Civic Responses to Violent Organized Crime in Mexico* (Woodrow Wilson Center's Mexico Institute and the Trans-Border Institute at the University of San Diego, 2014).

[103] Diana Negroponte, *Pillar IV of 'Beyond Merida:' Addressing the Socio-Economic Causes of Drug Related Crime and Violence in Mexico*, Woodrow Wilson Center's Mexico Institute, May 2011, available at http://www.wilsoncenter.org/sites/default/files/Merida%20-%20Pillar%20IV%20Working%20Paper%20Format1.pdf

[104] The State Department reprogrammed $8.5 million in FY2010 funding and $14 million in FY2011 funding to support pillar four projects. To date (FYs 2010, 2010 supplemental, 2011, 2012 and 2013), USAID has dedicated $35.9 million in Economic Support Fund monies to pillar four.

---

youth.[105] U.S.-funded pillar four activities were designed to complement the work of the Mexico's National Center for Crime Prevention and Citizen Participation, an entity within the Interior Department that implemented projects in high crime areas in 237 cities in 2012 where local authorities were making similar investments in crime prevention.

In support of this new strategy, USAID launched a three-year, $15 million Crime and Violence Prevention program in nine target communities identified by the Mexican government in Ciudad Juárez, Monterrey, Nuevo León, and Tijuana, Baja California. The program supports the development of community strategies to reduce crime and violence in the target localities, including outreach to at-risk youth, improved citizen-police collaboration, and partnerships with private sector enterprises. USAID also awarded $10 million in local grants to six civil society organizations for innovative crime prevention projects that engage at-risk youth and their families. USAID also supports a $1 million evaluation of crime in the target communities that will help the U.S. and Mexican governments understand the risk factors contributing directly to increased violence and enable both governments to identify successful models for replication.

Experiences in U.S. and Latin American cities have shown the importance that municipal-based crime prevention programs play in efforts to reduce violence. USAID has supported local prevention programs in Central America since the mid-2000s, and lessons learned can be drawn from that experience. Some may argue that similar programs in Mexico should be scaled up, while others may assert that Mexico, a middle-income country, has the capacity to pay for its own prevention programs. As a result, the U.S. government's pillar four programs were designed as pilots for future replication in other areas of Mexico with similar characteristics and vulnerabilities. Mexican participation and ultimate ownership and responsibility of these programs, as well as local civil society participation and oversight, will be crucial to sustaining these investments.

Pillar four appears to be a top priority for the Peña Nieto government and future bilateral efforts will likely seek to complement Mexico's National Crime and Violence Prevention Program. As previously stated, that program involves federal interventions in municipalities in 57 high crime areas. To bolster those interventions, USAID may expand the geographic areas in which it is supporting violence prevention programs. In addition, the State Department is supporting another key element of the program: drug demand reduction (DDR). For example, the DDR program which has already worked with the Organization of American States to develop a curriculum on drug treatment that has been given to 600 counselors in six states and supported research and clinical trials, is in the process of helping Mexico expand drug treatment courts throughout the country. As Mexico has made culture of lawfulness (CoL) education a required part of middle school curriculum, U.S. support has helped that curriculum reach more than 800,000 students during the 2013-2014 school year.[106]

---

[105] U.S. programs under the first objective may help refine the Mexican government's national crime prevention plan and support federal entities engaged in developing, monitoring, and evaluating municipal crime prevention efforts. Under the second objective, USAID may support the development and implementation of municipal crime prevention plans. Programs under the third objective may include helping communities build networks of resources for at-risk youth. See USAID-Mexico, "Pillar Four: Building Strong and Resilient Communities," http://www.usaid.gov/mx/pillariveng html.

[106] U.S. Embassy, February 2014.

Issues

Measuring the Success of the Mérida Initiative

With little publicly available information on what specific metrics the U.S. and Mexican governments are using to measure the impact of the Mérida Initiative, analysts have debated how bilateral efforts should be evaluated.[107] How one evaluates the Mérida Initiative largely depends on how one has defined the goals of the program. While the U.S. and Mexican governments' long-term goals for the Mérida Initiative may be similar, their short-term goals and priorities may be different. For example, both countries may strive to ultimately reduce the overarching threat posed by the DTOs—a national security threat to Mexico and an organized crime threat to the United States. However, their short-term goals may differ; Mexico may focus more on reducing drug trafficking-related crime and violence, while the United States may place more emphasis on aggressively capturing DTO leaders and seizing illicit drugs.

One basic measure by which Congress has evaluated the Mérida Initiative has been the pace of equipment deliveries and training opportunities. A December 2009 Government Accountability Office (GAO) report identified several factors that had slowed the pace of Mérida implementation.[108] It is unclear, though, whether more expeditious equipment deliveries to Mexico have resulted in a more positive evaluation of Mérida. Moreover, if equipment is not adequately maintained, its long-term impact could be reduced. Measures of the volume of training programs administered, including the number of individuals completing each course, have also been used to measure Mérida success. This measure is imperfect, however, as it does not capture the impact that a particular training course had on an individuals' performance. U.S. agencies are generally not currently measuring retention rates for those whom they have trained; some agencies have identified high turnover rates within the agencies as a major obstacle for the sustainability of Mérida-funded training programs.[109]

U.S.-funded antidrug programs in source and transit countries (of which Mexico is both) have also traditionally been evaluated by examining the number of DTO leaders arrested and the amount of drugs and other illicit items seized, along with the price and purity of drugs in the United States. Some analysts have attributed increased arrests and certain drug seizures (i.e., cocaine and methamphetamine) to success of the Mérida Initiative. Others have also highlighted the downward trend (since 2006) of cocaine availability and purity in the United States as evidence of the success of Mérida and other U.S.-funded antidrug efforts.

However, a principal challenge in assessing the success of Mérida is separating the results of those efforts funded via Mérida from those efforts funded through other border security and bilateral cooperation initiatives. The data available do not allow U.S. officials or analysts to determine the success that can be directly attributed to Mérida. Changes in seizure data and drug

---

[107] See, for example, Andrew Selee, *Success or Failure? Evaluating U.S.-Mexico Efforts to Address Organized Crime and Violence*, Center for Hemispheric Policy - Perspectives on the Americas Series, December 20, 2010.

[108] Government Accountability Office, *Status of Funds for the Mérida* Initiative, 10-253R, December 3, 2009.

[109] U.S. Agency for International Development, Justice Studies Center of the Americas, and Coordination Council for the Implementation of the Criminal Justice System and its Technical Secretariat (SETEC); Executive Summary of the General Report: Monitoring the Implementation of the Criminal Justice Reform in Chihuahua, the State of Mexico, Morelos, Oaxaca, and Zacatecas: 2007-2011, November 2012.

prices may not be directly related to U.S.-Mexican efforts to combat the DTOs. It is equally difficult to parcel out the reasons for periodic fluctuations in drug purity in the United States.

President Enrique Peña Nieto has vowed to reduce drug trafficking-related violence and crimes such as kidnapping and extortion. Should a decrease in drug trafficking-related deaths be used as an indicator of success for the Mérida Initiative? What if drug trafficking-related deaths continue to decline, but extortions and kidnappings continue to increase?

In addition to a decline in drug trafficking-related violence and crime, analysts have suggested that success in pillars two and four would be evidenced by, among other things, increases in popular trust in the police and courts. Measuring citizens' perceptions on crime and violence, on the one hand, as well as governmental effectiveness, on the other, could also prove useful.

Still others, including U.S. officials, have maintained that the success of the Mérida Initiative may be measured by a broad range of indicators that show increased bilateral cooperation. For instance, the State Department has cited the arrests and killings of high-profile DTO leaders that have been made since late 2009 as examples of the results of increased bilateral law enforcement cooperation.

Extraditions[110]

Another example of Mérida success—in the form of bilateral cooperation—cited by the State Department is the high number of extraditions from Mexico to the United States. Extraditions to the United States had started to increase under former President Vicente Fox (2000-2006), who like Felipe Calderón was from the opposition PAN, and reached 63 in 2006. Starting at 83 in 2007, in the six full years of the Calderón Administration, extraditions rose to nearly 100 a year Cooperation on extraditions peaked in 2012, the final year of the Calderón government, with 115 favorable responses to U.S. extradition requests, including 52 extraditions on drug-related offenses.

---

[110] June S. Beittel, Analyst in Latin American Affairs, contributed to this section.

Figure 2. Individuals Extradited from Mexico to the United States

1995-2013

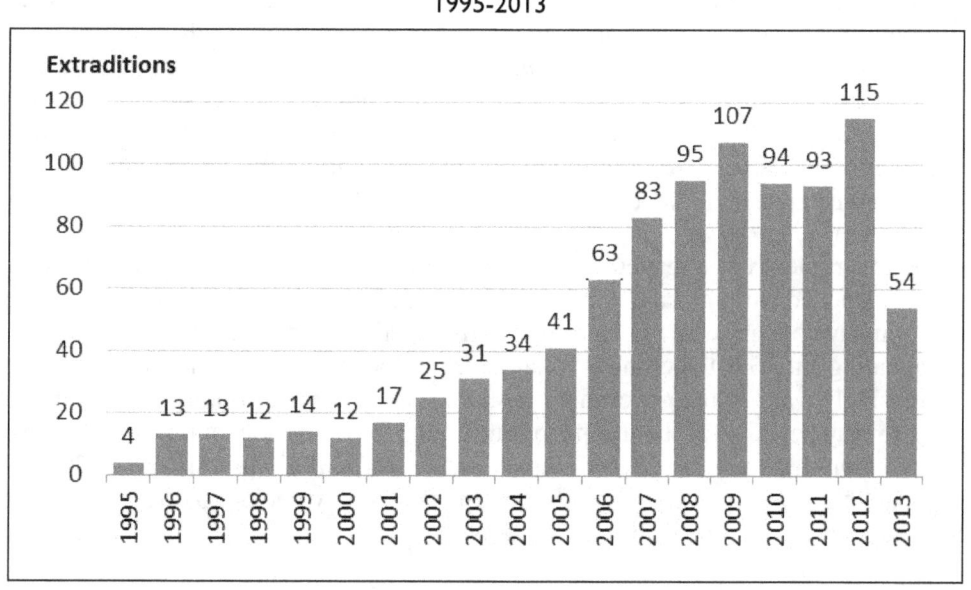

Sources: 1995-2006 data from U.S. Embassy to Mexico, U.S. - Mexico at a Glance: Law Enforcement at a Glance, http://www.usembassy-mexico.gov/eng/eataglance_law.html; 2007-2008 data from the Trans-Border Institute, Justice in Mexico, *News Report January 2009*, January 2009; 2009 data from the U.S. Department of State, "United States - Mexico Security Partnership: Progress and Impact," press release, May 19, 2010. 2010 and 2011 data from the U.S. Department of Justice. 2012-2013 data from the U.S. Department of State, *International Narcotics Control Strategy Report (INCSR)*, Vol. 1, March 2014.

In the first full year of President Enrique Peña Nieto's term in office (2013), however, the number of extraditions declined to 54. Given the transition to a new administration in Mexico, there are several possible reasons for that decline in extraditions. According to Mexican officials, extradition requests from the United States to Mexico declined from 108 in 2012 to 88 in 2013. Moreover, the Calderón government did not leave a large backlog of cases waiting to be processed in 2013.[111] In addition, the Mexican government may be attempting to show that the Mexican justice system, which is in the process of being reformed, is capable of arresting, trying, and convicting drug traffickers, including "El Chapo" Guzmán.[112] At a recent hearing, several Members of Congress vigorously urged State Department and DHS officials to seek Guzmán's extradition; those officials responded by stating that they are engaged in discussions with the Mexican government on that matter.[113]

Drug Production and Interdiction in Mexico

Drug eradication and alternative development programs have not been a focus of the Mérida Initiative even though Mexico is a major producer of cannabis (marijuana), opium poppy (used to produce heroin), and methamphetamine. According to U.S. government estimates, marijuana and

---

[111] CRS interview with Mexican official, March 13, 2014.

[112] Mark Stevenson, Alicia A. Caldwell, and Adriana Gomez Licon, "Drug Lord 'El Chapo' Guzman Charged in Mexico," *Associated Press*, February 24, 2014.

[113] U.S. Congress, House Committee on Homeland Security, *Taking Down the Cartels: Examining United States-Mexican Cooperation*, 113th Cong., 2nd sess., April 2, 2014.

opium poppy cultivation in rural Mexico expanded significantly in 2009-2010, before declining in 2011 due to drought conditions in crop-growing regions and slight increases in eradication. Eradication figures increased again slightly in 2012. No cultivation or eradication figures for Mexico appeared in the State Department's *International Narcotics Control Strategy Report* covering 2013, however press reports maintain that opium production has surged as cannabis production has fallen.[114] In addition, despite Mexican government import restrictions on precursor chemicals and efforts to seize precursor chemicals and dismantle clandestine labs, the production of methamphetamine appears to have continued at high levels.

The Mexican government has engaged its military in drug crop eradication efforts since the 1930s, but personnel constraints have inhibited recent eradication efforts. Indeed, increases in drug production have occurred as the government assigned more military forces to public security functions, including anti-DTO operations, than to drug crop eradication efforts. Should Mexicans become increasingly wary of the government's strategy of using the military to perform police functions, there may be calls for the troops to return to more traditional antidrug functions. Similarly, if drug production in Mexico further expands, particularly production of the potent and dangerous "black tar" variety of heroin, U.S. policy makers may decide to direct some Mérida assistance to support eradication efforts in Mexico.

The Mexican government has not traditionally provided support for alternative development even though many drug-producing regions of the country are impoverished rural areas where few licit employment opportunities exist. Alternative development programs have traditionally sought to provide positive incentives for farmers to abandon drug crop cultivation in lieu of farming other crops, but may be designed more broadly to assist any individuals who collaborate with DTOs out of economic necessity. In Colombia, studies have found that the combination of jointly implemented eradication, alternative development, and interdiction is more effective than the independent application of any one of these three strategies.[115] Despite those findings, alternative development often takes years to show results and requires a long-term commitment to promoting rural development, two factors which may lessen its appeal as a policy tool for Mexico.

While Mexico has made arresting drug kingpins a top priority, it has not given equal attention to the need to increase drug seizures. The State Department's *International Narcotics Control Strategy Reports* covering 2012 and 2013 assert that less than two percent of the cocaine estimated to transit Mexico is seized by Mexican authorities. Cocaine seizures in Nicaragua and other Central American countries often exceed Mexico's cocaine interdiction figures. Although the State Department has provided canines and NIIE equipment for interdiction at Mexico's borders and ports of entry, it does not appear to be a top priority of the Mérida Initiative.

---

[114] Nick Miroff, "Tracing the U.S. Heroin Surge Back South of the Border as Mexican Cannabis Output Falls, *Washington Post*, April 6, 2014.

[115] Vanda Felbab-Brown, Joel M. Jutkowitz, Sergio Rivas, et al. *Assessment of the Implementation of the United States Government's Support for Plan Colombia's Illicit Crop Reduction Components*, report produced for review by the U.S. Agency for International Development (USAID), April 17, 2009.

Human Rights Concerns and Conditions on Mérida Initiative Funding

There have been ongoing concerns about the human rights records of Mexico's military and police. The State Department's annual human rights reports covering Mexico have cited credible reports of police involvement in extrajudicial killings, kidnappings for ransom, and torture.[116] There has also been concern that the Mexican military has committed more human rights abuses since being tasked with carrying out public security functions. According to Mexico's Human Rights Commission (*Commisión Nacional de Derechos Humanos* or CNDH), complaints of human rights abuses by Mexico's Department of Defense (SEDENA) increased from 182 in 2006 to a peak of 1,800 in 2009 before falling since that time. Complaints of abuses against the Secretariat of the Navy (SEMAR) increased by 150% from 2010 to 2011 as its forces became more heavily involved in anti-DTO efforts, before decreasing in 2012 and 2013.[117] While troubling, only a small percentage of those allegations have resulted in the CNDH issuing recommendations for corrective action to SEDENA and SEMAR, which those agencies have largely accepted.[118]

In addition to expressing concerns about current abuses, Mexican and international human rights groups have criticized the Mexican government for failing to hold military and police officials accountable for past abuses. They fault the government for not taking further steps to comply with rulings by the Inter-American Court of Human Rights (IACHR) and decisions by Mexico's Supreme Court affirming that cases of military abuses against civilians should be tried in civilian courts. Since 2011, the military asserts that it has transferred "dozens of cases" to civilian jurisdiction, but information on those cases is not publicly available.[119]

Congress has expressed ongoing concerns about human rights conditions in Mexico. These concerns have intensified as U.S. security assistance to Mexico has increased under the Mérida Initiative. Congress has continued monitoring adherence to the "Leahy" vetting requirements that must be met under the Foreign Assistance Act (FAA) of 1961 as amended (22 U.S.C. 2378d)[120] and annual Department of Defense (DOD) appropriations[121] in order for Mexican security forces[122] to receive U.S. support.[123]

---

[116] U.S. Department of State, *Country Report on Human Rights Practices for 2012: Mexico*, April 2013.*Country Report: Mexico*, February 2014.

[117] These figures are from CNDH's annual activity reports. They are available in Spanish at http://www.cndh.org mx/Informes_Actividades

[118] In 2013, for example, the 811 complaints filed with CNDH against SEDENA resulted in 3 recommendations.

[119] For background, see Maureen Meyer, *Recent Developments on the Use of Military Jurisdiction in Mexico*, WOLA, January 31, 2012; *Country Report: Mexico*, February 2014.

[120] The codified Leahy law (22 U.S.C. 2378d) prohibits the furnishing of assistance authorized by the FAA and the Arms Export Control Act, as amended, (AECA) to any foreign security force unit that is credibly believed to have committed a gross violation of human rights.

[121] A provision in the annual DOD appropriations legislation prohibits the use of DOD funds to support any training program involving a unit of a foreign security or police force if the unit has committed a gross violation of human rights. P.L. 113-76 expands that prohibition to cover DOD equipment assistance programs as well.

[122] There is no FAA definition for the term "security force." DOD defines the term as "duly constituted military, paramilitary, police, and constabulary forces of a state."(DOD Dictionary of Military and Associated Terms, DOD Joint Publication 1-02, http://www.dtic.mil.)

[123] CRS Report R43361, *"Leahy Law" Human Rights Provisions and Security Assistance: Issue Overview*, coordinated (continued...)

Since FY2008, Congress has also conditioned U.S. assistance to the Mexican military and police on compliance with certain human rights standards. The FY2008 Supplemental Appropriations Act (P.L. 110-252), which provided the first tranche of Mérida funding, had less stringent human rights conditions than had been proposed earlier, largely due to Mexico's concerns that some of the conditions would violate its national sovereignty. The conditions required that 15% of INCLE and Foreign Military Financing (FMF) assistance be withheld until the Secretary of State reports in writing that Mexico is taking action in four human rights areas. Those were: 1) improving transparency and accountability of federal police forces; 2) establishing a mechanism for regular consultations among relevant Mexican government authorities, Mexican human rights organizations, and other relevant Mexican civil society organizations; 3) ensuring that civilian prosecutors and judicial authorities are investigating and prosecuting, in accordance with Mexican and international law, members of the federal police and military forces who have been credibly alleged to have committed violations of human rights, and the federal police and military forces are fully cooperating with the investigations; and 4) enforcing the prohibition on the use of testimony obtained through torture or other ill-treatment.

Similar human rights conditions were included in FY2009-FY2011 appropriations measures that funded the Mérida Initiative.[124] However, the first two conditions were not included in the 15% withholding requirement in the FY2012 Consolidated Appropriations Act (P.L. 112-74). The FY2013 Consolidated and Further Continuing Appropriations Act (P.L. 113-6) contained the same withholding requirement as P.L. 112-74.

Thus far, the State Department has submitted three 15% progress reports on Mexico to congressional appropriators (in August 2009, September 2010, and August 2012) that have met the statutory requirements for FY2008-FY2013 Mérida funds that had been on hold to be released. Nevertheless, the State Department has twice elected to hold back some funding pending further progress in key areas of concern. In the September 2010 report, for example, the State Department elected to hold back $26 million in FY2010 supplemental funds as a matter of policy until further progress was made in the areas of transparency and combating impunity. Those funds were not obligated until the fall of 2011.

In the August 2012 report, the State Department again decided to hold back all of the FY2012 funding that would have been subject to the conditions (roughly $18 million) as a matter of policy until it can work with Mexican authorities to determine steps to address key human rights challenges. Those include improving the ability of Mexico's civilian institutions to investigate

---

(...continued)

by Nina M. Serafino

[124] In P.L. 110-252, the human rights conditions applied to 15% of the funding for INCLE and FMF, or approximately $57 million dollars. In the FY2009 Omnibus Appropriations Act (P.L. 111-8), the 15% conditions applied all of the funding accounts but excluded amounts for judicial reform, institution building, anti-corruption and rule of law activities, which were earmarked at not less than $75 million. The total aid withheld was $33.4 million. In the FY2009 Supplemental (P.L. 111-32), the conditions effectively only applied to the $160 million in the INCLE account, or roughly $24 million, because the $260 million in FMF funds was excluded from the 15% withholding requirement. In the FY2010 Consolidated Appropriations Act (P.L. 111-117), the 15% withholding applied to all of the accounts but excluded assistance for judicial reform, institution building, anti-corruption and rule of law activities. The total aid withheld was some $12 million. In the FY2010 Supplemental Appropriations Act (P.L. 111-212), the conditions applied to 15% of the INCLE appropriated or roughly $26 million. The same conditions that were included in P.L. 111-117 applied to assistance provided in the FY2011 Department of Defense and Full-Year Continuing Appropriations Act (P.L. 112-10). According to the State Department, the FY2011 funds on hold totaled approximately $3.5 million. Email from State Department official, January 24, 2012.

and prosecute cases of human rights abuses; enhancing enforcement of prohibitions against torture and other mistreatment; and strengthening protection for human rights defenders.[125]

Human rights groups initially expressed satisfaction that President Peña Nieto had adopted a pro-human rights discourse and promulgated a law requiring state support for crime victims and their families. They have since been underwhelmed with his government's efforts to promote and protect human rights.[126] Some have therefore urged U.S. policymakers to closely monitor the Peña Nieto government's compliance with conditions on Mérida assistance and to continue rigorous vetting of Mexican individuals and units slated to receive U.S. training and equipment.[127] How the Peña Nieto government moves to improve the ability of Mexico's civilian institutions to investigate and prosecute cases of human rights abuses by security forces, enhance enforcement of prohibitions against torture and other mistreatment, and strengthen protection for human rights defenders, the media, migrants, and other vulnerable groups is likely to be closely scrutinized.

The State Department has established a high-level human rights dialogue with Mexico, provided human rights training for Mexican security forces, and implemented a number of human rights-related programs. In 2011, USAID launched a $5 million program being implemented by Freedom House to improve protections for Mexican journalists and human rights defenders. Human rights groups have acknowledged these efforts, but have criticized the U.S. government for failing to enforce Mérida's human rights restrictions and for backing Mexico's military-led security strategy.

Congress may choose to augment Mérida Initiative funding for human rights programs, such as ongoing training programs for military and police, or newer efforts, such as support for human rights organizations. Human rights conditions in Mexico, as well as compliance with conditions on Mérida assistance, are also likely to continue to be important oversight issues. The FY2014 Consolidated Appropriations Act (P.L. 113-76) includes several human rights provisions regarding aid to Mexico. Those provisions withhold 15% of assistance to the Mexican military and police until the State Department reports that progress has been made in meeting four human rights conditions.[128] They also appear to require a report from the State Department within 60 days of the measure's enactment (January 17, 2014) on progress made in meeting the human rights conditions in the FY2012 and FY2013 appropriations laws (P.L. 112-74 and P.L. 113-6).[129]

---

[125] U.S. Department of State, *Mexico - Merida Initiative Report ("15 Percent" Report)*, August 30, 2012.

[126] José Miguel Vivanco, *Mexico: President's Disappointing First Year on Human Rights*, Human Rights Watch, November 26, 2013.

[127] Restrictions on certain aid to Mexico's military and police have been included in each of the Mérida appropriations measures since P.L. 110-252. See CRS Report R41349, *U.S.-Mexican Security Cooperation: The Mérida Initiative and Beyond*.

[128] Those conditions require the Secretary of State to report that the Mexican government (1) has reformed its military justice system to require that military abuses against civilians are investigated and prosecuted in the civilian justice system; (2) is enforcing prohibitions against torture and the use of testimony obtained through torture; (3) is ensuring that military and police are immediately transferring detainees to the custody of civilian judicial authorities and are cooperating with such authorities in such cases; and, (4) is searching for the victims of enforced and involuntary disappearances and prosecuting those responsible for such crimes. They are outlined in S.Rept. 113-81 accompanying the Senate version of the FY2014 State-Foreign Operations appropriations bill (S. 1372).

[129] The reporting requirement originally appeared in H.Rept. 113-185 accompanying the House Appropriations Committee's version of the FY2014 State-Foreign Operations appropriations bill, H.R. 2855.

Role of the U.S. Department Of Defense in Mexico

In contrast to Plan Colombia, the Mérida Initiative does not include an active U.S. military presence in Mexico, largely due to Mexican concerns about national sovereignty stemming from past conflicts with the United States. The Department of Defense (DOD) did not play a primary role in designing the Mérida Initiative and is not providing assistance through Mérida aid accounts. However, DOD administered assistance provided through the Foreign Military Financing (FMF) account, which was part of Mérida until FY2012. As an implementing agency, DOD's role largely involved overseeing the procurement and delivery of Mérida equipment (including 15 aircraft and 13 ion scanners) for Mexican security forces.

Despite DOD's limited role in the Mérida Initiative, military cooperation between the two countries has been increasing, as have DOD training and equipment programs to support the Mexican military.[130] For example, according to press reports, in response to a request from the Mexican government, DOD began sending unmanned aerial vehicles into Mexico to gather intelligence on criminal organizations.[131] As previously mentioned ("Mexico's Southern Borders"), DOD is providing training and equipment to Mexican military forces patrolling the country's southern borders. More broadly, DOD assistance aims to support efforts to improve security in high-crime areas, track and capture DTO operatives, strengthen border security, and disrupt flows of illicit finances, drugs, and other materials.

There are a variety of funding streams that support DOD training and equipment programs. Some DOD training programs are funded by annual State Department appropriations for FMF, which totaled $8 million in FY2011 and $7 million in FY2012 and FY2013, and International Military Education and Training (IMET), which has exceeded $1 million per year for the past several years (see **Appendix A**). Apart from the Mérida Initiative and other State Department funding, DOD has its own legislative authorities to provide certain counterdrug assistance. DOD programs in Mexico are overseen by the U.S. Northern Command (NORTHCOM), which is located at Peterson Air Force Base in Colorado. DOD can provide counterdrug assistance under guidelines outlined in Section 1004 of P.L. 101-510, as amended, and can provide additional assistance to certain countries as provided for in Section 1033 of P.L. 105-85, as amended. DOD 1004 and 1003 support to Mexico totaled roughly $61.5 million in FY2011, $84.4 million in FY2012, and $68.8 million in FY2013.[132]

The aforementioned counternarcotics funding has enabled NORTHCOM to train and equip an increasing number of Mexican military personnel. In FY2013, for example, NORTHCOM trained more than 3,000 military personnel, a 44% increase over FY2011. Training included courses on "information fusion, surveillance, interdiction, cyber security, logistics, and professional development."[133] Equipping efforts provided non-lethal equipment (such as communications tools, aircraft modifications, night vision, boats, etc.) to support those training courses.

---

[130] Roderic Ai Camp, *Armed Forces and Drugs: Public Perceptions and Institutional Challenges*, Woodrow Wilson Center, Working Paper Series on U.S.-Mexico Security Collaboration, May 2010; Richard D. Downie, "Critical Decisions in Mexico: the Future of U.S./Mexican Defense Relations," *Strategic Issues in U.S.-Latin American Relations*, vol. 1, no. 1 (July 2011).

[131] Ginger Thompson and Mark Mazzetti, "U.S. Drones Fly Deep in Mexico to Fight Drugs," *New York Times*, March 16, 2011.

[132] Electronic correspondence with Northern Command legislative affairs, April 1, 2014.

[133] Electronic correspondence with Northern Command legislative affairs, March 31, 2014.

Since DOD counterdrug assistance is obligated out of global accounts and the agency is not required to submit country-specific requests to Congress for its programs, obtaining recent data on DOD programs and plans for Mexico may be difficult. Regardless, policy makers may want to receive periodic briefings on those efforts in order to guarantee that current and future DOD programs are being adequately coordinated with Mérida Initiative efforts. They may also want to ensure that DOD-funded programs are not inadvertently reinforcing the militarization of public security in Mexico. Experts have urged the United States "not to focus too much on military assistance and neglect other, more effective forms of aid ... [such as assistance for] the development, training, and professionalization of Mexico's law enforcement officers."[134]

Balancing Assistance to Mexico with Support for Southwest Border Initiatives

The Mérida Initiative was designed to complement domestic efforts to combat drug demand, drug trafficking, weapons smuggling, and money laundering. These domestic counter-drug initiatives are funded through regular and supplemental appropriations for a variety of U.S. domestic agencies. As the strategy underpinning the Mérida Initiative has expanded to include efforts to build a more modern border (pillar three) and to strengthen border communities (pillar four), policy makers may consider how best to balance the amount of funding provided to Mexico with support for related domestic initiatives.[135]

Regarding support for law enforcement efforts, some would argue that there needs to be more federal support for states and localities on the U.S. side of the border that are dealing with crime and violence originating in Mexico. Of those who endorse that point of view, some are encouraged that the Obama Administration has increased manpower and technology along the border, whereas others maintain that the Administration's efforts have been insufficient to secure the border.[136] In contrast, some maintain that it is impossible to combat transnational criminal enterprises by solely focused on the U.S. side of the border, and that domestic programs must be accompanied by continued efforts to build the capacity of Mexican law enforcement officials. They maintain that if recent U.S. efforts are perceived as an attempt to "militarize" the border, they may damage U.S.-Mexican relations and hinder bilateral security cooperation efforts. Mexican officials from across the political spectrum have expressed concerns about the construction of border fencing and the effects of border enforcement on migrant deaths.[137]

With respect to pillar four of the updated strategy, as previously mentioned, Mexico and the United States are supporting programs to strengthen communities in Ciudad Juárez, Monterrey, and Tijuana. In targeting those communities most affected by the violence, greater efforts will necessarily be placed on community-building in Ciudad Juárez and Tijuana than on their sister cities in the United States. However, if the U.S. government provides aid to these communities in

---

[134] Robert C. Bonner, "The New Cocaine Cowboys: How to Defeat Mexico's Drug Cartels," *Foreign Affairs*, vol. 89, no. 4 (July/August 2010).

[135] The SWBCS, 2011 includes a new chapter on U.S. efforts to promote strong communities by, in part, increasing crime prevention efforts and drug prevention and treatment programs in U.S. border communities.

[136] For a fuller discussion of U.S. border enforcement efforts, see CRS Report R42138, *Border Security: Immigration Enforcement Between Ports of Entry*, by Lisa Seghetti.

[137] See, for example, Marc R. Rosenblum, *Obstacles and Opportunities for Regional Cooperation: the U.S.-Mexico Case*, Migration Policy Institute, April 2011.

Mexico, some may argue that there should also be federal support for the adjacent U.S. border cities. For example, initiatives aimed at providing youth with education, employment, and social outlets might reduce the allure of joining a DTO or local gang. Some may contend that increasing these services on the U.S. side of the border as well as the Mexican side could be beneficial.

Integrating Counterdrug Programs in the Western Hemisphere

U.S. State Department-funded counterdrug assistance programs in the Western Hemisphere are currently in transition. Counterdrug assistance to Colombia and the Andean region is in decline after record assistance levels that began with U.S. support for Plan Colombia in FY2000 and peaked in the mid-2000s. Antidrug aid to Mexico increased dramatically in FY2008-FY2010 as a result of the Mérida Initiative, but is also now gradually being reduced. Conversely, funding for Central America is increasing as a result of the Central American Regional Security Initiative (CARSI)[138] and support for the Caribbean increased in FY2010 and has remained relatively stable due to the Caribbean Basin Security Initiative (CBSI).

The Obama Administration has taken steps to coordinate the aforementioned country and regional antidrug programs and to ensure that U.S.-funded efforts complement the efforts of partner governments and other donors. The Administration has appointed a coordinator within the State Department (the Principal Deputy Assistant Secretary of State for Western Hemisphere Affairs) to oversee the planning and implementation of the aforementioned security assistance packages. The Office of National Drug Control Policy (ONDCP) and the National Security Council conduct annual reviews of counterdrug efforts in the Americas. ONDCP and the State Department use a high-level committee process to oversee programming and planning. The Administration is encouraging countries that have received U.S. assistance in the past—particularly Colombia—to share technical expertise with other countries in the region, a strategy that analysts have recommended. One area in which closer cooperation between the United States, partner governments, and other donors will likely be necessary is in efforts to better secure the porous Mexico-Guatemala and Mexico-Belize borders.

Outlook

Mexico has experienced a transition from a PAN Administration focused on combating organized crime to a PRI government focused on bolstering competitiveness by enacting structural reforms. As a result, security issues may take a back seat to economic issues on the bilateral agenda for the first time since September 2001. On May 2, 2013, President Obama traveled to Mexico for a trip focused on enhancing economic cooperation and expanding educational exchanges between the two countries.[139] When asked about changes in Mexico's security strategy, President Obama said "it is up to the Mexican people to determine their security structures and how [they will engage] with other nations, including the United States."[140] He reiterated his Administration's support for Mexico's efforts to reduce violence and criminality, including continued U.S. assistance.

---

[138] CRS Report R41731, *Central America Regional Security Initiative: Background and Policy Issues for Congress*, by Peter J. Meyer and Clare Ribando Seelke.

[139] The White House, Office of the Press Secretary, "White House Fact Sheet on U.S.-Mexico Partnership," Press Release, May 2, 2013.

[140] The White House, Office of the Press Secretary, "Remarks by President Obama and President Peña Nieto of Mexico (continued...)

When examining the future of the Mérida Initiative, Congress may first consider defining the desired end state of the Mérida Initiative. Congress may then seek to ensure that those who are implementing the Initiative have developed adequate metrics to measure progress, and that those metrics are shared with Congress for review and oversight. Given the level of progress that has been made thus far, the current Mérida strategy may be deemed sufficient or insufficient. If it is judged insufficient, Congress may consider how it might be improved. When considering future assistance for the Mérida Initiative, Congress may compare how much funding programs in Mexico, an upper middle income country, are receiving from the Peña Nieto government, and whether U.S. funding is complementing or duplicating Mexican efforts.

(...continued)

in a Joint Press Conference," Press Release, May 2, 2013.

Appendix A. U.S. Assistance to Mexico

Table A-1. U.S. Assistance to Mexico by Account, FY2007-FY2014

(U.S. $ millions)

Account	FY2007	FY2008[a]	FY2009[b]	FY2010	FY2011	FY2012	FY2013	FY2014 (est.)	FY2015 (req.)
INCLE	36.7	242.1	454.0[c]	365.0[d]	117.0	248.5	195.1	148.1	80.0
ESF	11.4	34.7	15.0	15.0	18.0	33.3	32.1	46.1	35.0
FMF	0.0	116.5	299.0[e]	5.3	8.0	7.0	6.6	7.0	5.0
IMET	0.1	0.4	0.8	1.0	1.0	1.0	1.2	1.4	1.5
NADR	1.3	1.4	3.9	3.9	5.7	5.4	3.8	3.9	2.9
GHCS[f]	3.7	2.7	2.9	3.5	3.5	1.0	0.0	0.0	0.0
DA	12.3	8.2	11.2	10.0	25.0	33.4	26.2	0.0	12.5
TOTAL	65.4	405.9	786.8	403.7	178.2	329.6	265.0	206.5	136.9

Sources: U.S. Department of State, *Congressional Budget Justification for Foreign Operations FY2008-FY2015.*

Notes: GHCS=Global Health and Child Survival; DA=Development Assistance; ESF=Economic Support Fund; FMF=Foreign Military Financing; IMET=International Military Education and Training; INCLE=International Narcotics Control and Law Enforcement; NADR=Non-proliferation, Anti-terrorism and Related Programs. Funds are accounted for in the fiscal year for which they were appropriated as noted below:

a. FY2008 assistance includes funding from the Supplemental Appropriations Act, 2008 (P.L. 110-252).

b. FY2009 assistance includes FY2009 bridge funding from the Supplemental Appropriations Act, 2008 (P.L. 110-252) and funding from the Supplemental Appropriations Act, 2009 (P.L. 111-32).

c. $94 million provided under P.L. 111-32 and counted here as part of FY2009 funding was considered by appropriators "forward funding" intended to address in advance a portion of the FY2010 request.

d. $175 million provided in the FY2010 supplemental (P.L. 111-212) and counted here as FY2010 funding was considered by appropriators as "forward funding" intended to address in advance a portion of the FY2011 request.

e. $260 million provided under a FY2009 supplemental (P.L. 111-32) and counted here as FY2009 funding was considered by appropriators "forward funding" intended to address in advance a portion of the FY2010 request.

f. Prior to FY2008, the Global Health and Child Survival account was known as Child Survival and Health.

Appendix B. U.S. Domestic Efforts to Complement the Mérida Initiative

Drug Demand

Drug demand in the United States fuels a multi-billion dollar illicit industry. In 2012, about 23.9 million individuals were current (past month) illegal drug users, representing 9.2% of individuals aged 12 and older.[141] Administration officials and experts alike have acknowledged that U.S. domestic demand for illegal drugs is a significant factor driving the global drug trade, including the drug trafficking-related crime and violence that is occurring in Mexico and other source and transit countries.[142]

In April 2013, the Administration released its 2013 National Drug Control Strategy, which continues to emphasize the need to reduce U.S. drug demand. The Strategy furthers the goal of cutting drug use among youth by 15% by 2015.[143] Drug policy experts have praised the Administration's focus on reducing consumption, but criticized the Administration for requesting relatively modest budget increases in funding for treatment programs.[144] Some have questioned whether the federal government allocates enough of the drug budget to adequately address the demand side; the most recent drug budget continues to spend a majority of funds on supply reduction programs including drug crop eradication in source countries, interdiction, and domestic law enforcement efforts.[145] In addition to federal efforts, however, many state, local, and nonprofit agencies also channel funds toward demand reduction.

Firearms Trafficking[146]

Illegal firearms trafficking from the United States has been cited as a significant factor in the drug trafficking-related violence in Mexico. To address this issue, the Bureau of Alcohol, Tobacco, Firearms, and Explosives (ATF) stepped up enforcement of domestic gun control laws in the four Southwest border states under an agency-wide program known as "Project Gunrunner." ATF has

---

[141] See the National Survey on Drug Use and Health, an annual survey of approximately 67,500 people, including residents of households, non-institutionalized group quarters, and civilians living on military bases. The survey is administered by the Substance Abuse and Mental Health Services Administration of the U.S. Department of Health and Human Services.

[142] See, for example, testimony of R. Gil Kerlikowske, Director, Office of National Drug Control Policy, before the U.S. Congress, House Committee on Oversight and Government Reform, Subcommittee on National Security and Foreign Affairs, *Transnational Drug Enterprises (Part II): U.S. Government Perspectives on the Threat to Global Stability and U.S. National Security*, 111th Cong., 2nd sess., March 30, 2010.

[143] That strategy is available at http://www.whitehouse.gov/ondcp/2013-national-drug-control-strategy . For more information on the National Drug Control Strategy and the Office of National Drug Control Policy (ONDCP), see CRS Report R41535, *Reauthorizing the Office of National Drug Control Policy: Issues for Consideration*, by Lisa N. Sacco and Kristin Finklea.

[144] See, for example, Testimony of John T. Carnevale, President, Carnevale Associates, before the House Oversight and Government Reform Subcommitee on Domestic Policy, April 14, 2010.

[145] Office of National Drug Control Policy, *FY2013 Budget and Performance Summary*, April 2012.

[146] For background, see archived CRS Report R40733, *Gun Trafficking and the Southwest Border*, by Vivian S. Chu and William J. Krouse as well as CRS Report R42987, *Gun Control Legislation in the 113th Congress*, by William J. Krouse.

---

also trained Mexican law enforcement officials to use its electronic tracing (eTrace) program, through which investigators are sometimes able to trace the commercial trail and origin of recovered firearms. In the past, ATF has periodically released data on firearms traces performed for Mexican authorities. Although substantive methodological limitations preclude using trace data as a proxy for the larger population of "crime guns" in Mexico or the United States, trace data have proven to be a useful indicator of trafficking trends and patterns. In June 2009, GAO recommended to the Attorney General that he should direct ATF to update regularly its reporting on aggregate firearms trace data and trends.[147]

In February 2011, ATF came under intense congressional scrutiny for a Phoenix, AZ-based Project Gunrunner investigation known as Operation Fast and Furious, when ATF whistleblowers reported that suspected straw purchasers[148] had been allowed to acquire relatively large quantities of firearms as part of long-term gun trafficking investigations.[149] Some of these firearms are alleged to have "walked," or been trafficked to gunrunners and other criminals, before ATF moved to arrest the suspects and seize all of their contraband firearms. Two of those firearms were reportedly found at the scene of a shootout near the U.S.-Mexico border where U.S. Border Patrol Agent Brian Terry was shot to death.[150] Questions have also been raised about whether a firearm that was reportedly used to murder ICE Special Agent Jamie Zapata and wound Special Agent Victor Avila in Mexico on February 15, 2011, was initially trafficked by a subject of a Houston, TX-based Project Gunrunner investigation.[151] While it remains an open question whether ATF or other federal agents were in a position to interdict the firearms used in these deadly attacks before they were smuggled into Mexico,[152] neither DOJ nor ATF informed their Mexican counterparts about these investigations and the possibility that some of these firearms could be reaching Mexico.[153]

Legislators in both the United States and Mexico have voiced ongoing concerns about Operation Fast and Furious.[154] Repeated congressional inquiries prompted U.S. Attorney General Eric Holder to direct his Inspector General to conduct a third evaluation of Project Gunrunner, which was delivered to Congress in September 2012.[155]

---

[147] GAO, *Firearms Trafficking: U.S. Efforts to Combat Arms Trafficking to Mexico Face Planning and Coordination Challenges*, GAO-09-709, June 2009, p. 59.

[148] A "straw purchase" occurs when an individual poses as the actual transferee, but he is actually acquiring the firearm for another person. In effect, he serves as an illegal middleman. Straw purchases can be prosecuted under two provisions of the Gun Control Act of 1968, as amended (18 U.S.C. 922(a)(6) and 18 U.S.C. §924(a)(1)(A)).

[149] James v. Grimaldi and Sari Horwitz, "ATF Probe Strategy Is Questioned," *Washington Post*, February 2, 2011.

[150] Ibid.

[151] Ibid.

[152] Operation Fast and Furious was launched in November 2009. It was approved as an Organized Crime and Drug Enforcement Task Force (OCDETF) investigation in February 2010. As an OCDETF investigation, it was then directed largely by the U.S. Attorney's Office in Phoenix. While ICE and Internal Revenue Service (IRS) agents were also part of this investigation, so far their role in this operation has not generated public or congressional scrutiny.

[153] Richard A. Serrano, "U.S. Embassy Kept in Dark as Guns Flooded Mexico," *Salt Lake Tribune*, July 25, 2011.

[154] Dennis Wagner, "Gun Shop Told ATF Sting Was Perilous," *Arizona Republic*, April 15, 2011, p. A1.

[155] United States Department of Justice, Office of the Inspector General, *Statement of Michael E. Horowitz, Inspector General, U.S. Department of Justice before the House Committee on Oversight and Government Reform Concerning Report by the Office of the Inspector General on the Review of ATF's Operation Fast and Furious and Related Matters*, September 20, 2012, http://www.justice.gov/oig/testimony/t1220.pdf.

Separately, in July 2011, the Office of Management and Budget (OMB) approved an ATF multiple rifle sales reporting requirement for a three-year period.[156] Under this reporting requirement, federally licensed gun dealers in Southwest border states are required to report to ATF whenever they make multiple sales or other dispositions of more than one rifle within five consecutive business days to an unlicensed person.[157]

Bulk Cash Smuggling/Money Laundering

It is estimated that between $19 billion and $29 billion in illicit proceeds flow from the United States to drug trafficking organizations and other organized criminal groups in Mexico each year.[158] Much of this money is generated from the illegal sale of drugs in the United States. In their efforts to transfer illegal earnings from the United States to Mexico, and in the process make this cash appear legitimate, Mexican criminal networks rely upon a number of money laundering techniques, including smuggling bulk cash across the Southwest border; exploiting traditional banking and money services businesses (MSBs); using trade-based money laundering schemes; abusing prepaid access devices (also referred to as stored value cards); and leveraging mobile and electronic payment systems such as smartphones.[159] The illicit proceeds may then be used by DTOs and other criminal groups to acquire weapons in the United States and to corrupt law enforcement and other public officials.

Bulk cash smuggling—the physical movement of cash across the border—has been a prominent means by which criminals move illegal profits from the United States into Mexico. A number of federal law enforcement agencies, including CBP and ICE are involved in U.S. efforts to stem bulk cash smuggling. In 2005, ICE and CBP launched a program known as "Operation Firewall" that increased investigative efforts against bulk cash smuggling in the U.S.-Mexico border region as well as elsewhere. At the end of May 2012, the operation had yielded "more than 6,700 seizures totaling more than $621 million, and arrests of over 1,400 individuals. These efforts include[d] 480 international seizures totaling more than $271 million and 310 international arrests."[160]

---

[156] Office of Management and Budget, Office of Information and Regulatory Affairs, Reviews Completed in the Last 30 Days, DOJ-ATF, Report of Multiple Sale or Other Disposition of Certain Semi-Automatic Rifles, OMB Control Number: 1140-0100.

[157] This reporting requirement is limited to firearms that are (1) semiautomatic, (2) chambered for ammunition of greater than .22 caliber, and (3) capable of accepting a detachable magazine.

[158] DHS, *United States-Mexico Bi-National Criminal Proceeds Study*, June 2010.

[159] Celina B. Realuyo, *It's All about the Money: Advancing Anti-Money Laundering Efforts in the U.S. and Mexico to Combat Transnational Organized Crime*, Woodrow Wilson Center Mexico Institute, May 2012, pp. 6-13. See also Douglas Farah, *Money Laundering and Bulk Cash Smuggling: Challenges for the Merida Initiative*, Woodrow Wilson Center's Mexico Institute, Working Paper Series on U.S.-Mexico Security Cooperation, May 2010. Countering financial crimes—including money laundering and bulk cash smuggling—is one effort outlined by the National Southwest Border Counternarcotics Strategy (SWBCS). To curb the southbound flow of money from the sale of illicit drugs in the United States, the SWBCS includes several goals: stemming the flow of southbound bulk cash smuggling, prosecuting the illegal use of MSBs and electronic payment devices, increasing targeted financial sanctions, enhancing multilateral/bi-national collaboration, and empirically assessing the money laundering threat. ONDCP, *National Southwest Border Counternarcotics Strategy*, 2011.

[160] Statement for the Record, Leigh H. Winchell, then-Assistant Director for Programs, Homeland Security Investigations, Immigration and Customs Enforcement, in U.S. Congress, Senate Committee on Homeland Security and Governmental Affairs, Permanent Subcommittee on Investigations, *U.S. Vulnerabilities to Money Laundering, Drugs, and Terrorist Financing: HSBC Case History*, 112th Cong., 2nd sess., July 17, 2012, S.Hrg. 112–597 (Washington: GPO, 2012), p. 14. Hereafter Winchell, Statement for the Record.

As noted, criminal networks can take advantage of traditional banks and MSBs (such as money transfer companies) with lax anti-money laundering programs. In December 2012, DOJ announced that banking giant HSBC had agreed to pay $1.3 billion for "failing to maintain an effective anti-money laundering program and to conduct appropriate due diligence on its foreign correspondent account holders," among other charges.[161] HSBC allowed more than $881 million tied to the Sinaloa criminal network in Mexico and the Norte del Valle Cartel in Colombia to flow through the U.S. banking system.

Trade-based money laundering involves criminals "disguising the proceeds of crime and moving value through the use of trade transactions in attempt to legitimise their illicit origins."[162] To facilitate international investigations of trade-based money laundering networks, ICE developed the Trade Transparency Unit (TTU) model in 2004. ICE notes that "[o]ne of the most effectivve ways to identify instances and patterns of trade-based money laundering is through the exchange and subsequent analysis of trade data for anomalies that would only be apparent by examining both sides of a trade transaction."[163] ICE has a TTU in Mexico as well as others in Argentina, Brazil, Colombia, Paraguay, and Panama.[164]

Criminal networks have also turned to stored value cards to move money generated by their illicit activities. With these cards, criminals are able to avoid the reporting requirement under which they would have to declare any amount over $10,000 in cash moving across the border. Current federal regulations regarding international transportation only apply to *monetary instruments* as defined under the Bank Secrecy Act.[165] Of note, stored value cards are not considered monetary instruments under current law. The Financial Crimes Enforcement Network (FinCEN)[166] has issued a final rule, defining "stored value" as "prepaid access" and implementing regulations regarding the recordkeeping and suspicious activity reporting requirements for prepaid access products and services.[167] This rule does not, however, directly address whether stored value or prepaid access cards would be subject to current regulations regarding the international transportation of monetary instruments. A separate proposed rule would amend the definition of "monetary instrument," for the purposes of BSA international monetary transport regulations, to include prepaid access devices.[168] Even if FinCEN were to issue a final rule and implement

---

[161] Department of Justice, "HSBC Holdings Plc. and HSBC Bank USA N.A. Admit to Anti-Money Laundering and Sanctions Violations, Forfeit $1.256 Billion in Deferred Prosecution Agreement," December 11, 2012.

[162] Financial Action Task Force, "Trade Based Money Laundering," June 23, 2006, p. 1. Techniques that facilitate this form of money laundering include over- and under- invoicing (and over- and under- shipping) or multiple invoicing of goods and services as well as falsely describing goods and services.

[163] See http://www.ice.gov/trade-transparency/.

[164] Ibid.

[165] 31 U.S.C. §5312 defines a monetary instrument as "(A) United States coins and currency; (B) as the Secretary may prescribe by regulation, coins and currency of a foreign country, travelers' checks, bearer negotiable instruments, bearer investment securities, bearer securities, stock on which title is passed on delivery, and similar material; and (C) as the Secretary of the Treasury shall provide by regulation for purposes of sections 5316 and 5331 , checks, drafts, notes, money orders, and other similar instruments which are drawn on or by a foreign financial institution and are not in bearer form."

[166] FinCEN, under the Department of the Treasury, administers the BSA and the nation's financial intelligence unit. FinCEN also supports law enforcement, intelligence, and regulatory agencies by analyzing and sharing financial intelligence information. For more information, see http://www.fincen.gov/about_fincen/wwd/strategic.html.

[167] Department of the Treasury, Financial Crimes Enforcement Network, "Bank Secrecy Act Regulations—Definitions and Other Regulations Relating to Prepaid Access," 76, No. 146 *Federal Register* 45403-45420, July 29, 2011.

[168] Department of the Treasury, "Bank Secrecy Act Regulations Definition of "Monetary Instrument," 76 *Federal Register* 64049, October 17, 2011. Entities such as the Senate Caucus on International Narcotics Control have urged the (continued...)

---

regulations requiring individuals leaving the United States to declare stored value, the GAO has identified several challenges that would remain.[169] These challenges relate to law enforcement's ability to detect the actual cards and to differentiate legitimate from illegitimate stored value on cards; travelers' abilities to remember the amount of stored value on any given card; and law enforcement's ability to determine where illegitimate stored value is physically held and subsequently freeze and seize the assets.

In addition to prepaid access devices, Mexican criminal networks can launder money via digital currency accounts, e-businesses that facilitate money transfers via the Internet, and mobile payment systems wherein traffickers have remote access—often via cell phones—to hard-to-trace funds.[170] The current extent to which criminals may abuse mobile banking and web-based transactions is unknown, but experts are considering this as an "emerging avenue for cross-border money movements and laundering."[171]

U.S. efforts against money laundering and bulk cash smuggling are increasingly moving beyond the federal level as well, as experts have recommended.[172] In December 2009, for example, ICE opened a bulk cash smuggling center to assist U.S. federal, state, and local law enforcement agencies track and disrupt illicit funding flows. The United States and Mexico have created a Bilateral Money Laundering Working Group to coordinate the investigation and prosecution of money laundering and bulk cash smuggling. A Bi-national Criminal Proceeds Study revealed a number of major points along the Southwest border where bulk cash is smuggled.[173] Information provided from studies such as these may help inform policy makers and federal law enforcement personnel and assist in their decisions regarding where to direct future efforts against money laundering.

(...continued)

Administration to finalize this rule. See, for instance, Senate Caucus on International Narcotics Control, *The Buck Stops Here: Improving U.S. Anti-Money Laundering Practices,* April 2013.

[169] GAO, *Moving Illegal Proceeds: Challenges Exist in the Federal Government's Effort to Stem Cross Border Smuggling,* October 2010, pp. 48–49.

[170] Douglas Farah, *Money Laundering and Bulk Cash Smuggling: Challenges for the Merida Initiative,* Woodrow Wilson Center Mexico Institute, Working Paper Series on U.S.-Mexico Security Cooperation, May 2010, p. 161.

[171] Celina B. Realuyo, *It's All about the Money: Advancing Anti-Money Laundering Efforts in the U.S. and Mexico to Combat Transnational Organized Crime,* Woodrow Wilson Center Mexico Institute, May 2012, p. 13.

[172] Farah, op. cit.

[173] DHS, *United States - Mexico Bi-National Criminal Proceeds Study,* 2010.

Appendix C. Selected U.S.—Mexican Law Enforcement Partnerships

Border Enforcement Security Task Forces (BEST)

The BEST Initiative is a multi-agency initiative, led by Immigration and Customs Enforcement (ICE) within the Department of Homeland Security (DHS), wherein task forces seek to identify, disrupt, and dismantle criminal organizations posing significant threats to border security—both along the Southwest border with Mexico as well as along the Northern border with Canada.[174] Through the BEST Initiative, ICE partners with the U.S. Customs and Border Protection (CBP); the Drug Enforcement Administration (DEA); the Bureau of Alcohol, Tobacco, Firearms and Explosives (ATF); the Federal Bureau of Investigation (FBI); U.S. Coast Guard; and U.S. Attorneys' Offices; as well as local, state, and international law enforcement agencies. In particular, the Mexican Secretariat for Public Security (SSP) or federal police has been a partner along the Southwest border. There are currently 35 BEST teams around the country. BEST is the umbrella for activities such as the Vetted Arms Trafficking Group and the ICE Border Liaison Program.

Operation Against Smugglers (and Traffickers) Initiative on Safety and Security (OASISS)

CBP and the Mexican government have partnered through OASISS, a bi-lateral program aimed at enhancing both countries' abilities to prosecute alien smugglers and human traffickers along the Southwest border. Through OASISS, which has been in place since 2005, the Mexican government is able to prosecute alien smugglers apprehended in the United States.[175] This program is supported by the Border Patrol International Liaison Unit, which is responsible for establishing and maintaining working relationships with foreign counterparts in order to enhance border security.

Illegal Drug Program (IDP)

The Illegal Drug Program (IDP) is an agreement between ICE and the Mexican Attorney General's Office (PGR) wherein ICE can transfer cases of Mexican nationals smuggling drugs into the United States to the PGR for prosecution.[176] The program was initiated in Nogales, AZ, in October 2009 and subsequently adopted in El Paso, TX. Under the IDP, the U.S. Attorneys'

---

[174] Department of Homeland Security, U.S. Immigration and Customs Enforcement, *Border Enforcement Security Task Forces.*

[175] Testimony by Allen Gina, Assistant Commissioner, Office of Intelligence and Operations Coordination, U.S. Customs and Border Protection, Department of Homeland Security before the U.S. Congress, House Committee on Homeland Security, Subcommittee on Border, Maritime, and Global Counterterrorism, and House Committee on Foreign Affairs, Subcommittee on Western Hemisphere, *U.S.-Mexico Security Cooperation: Next Steps for the Merida Initiative*, 111th Cong., 1st sess., May 27, 2010.

[176] For more information on the IDP, see U.S. Immigration and Customs Enforcement, "ICE, Mexican authorities meet and agree to prosecution plan for drug smugglers captured at the border: DHS, Government of Mexico announce new agreement to help curb narcotics smuggling," press release, April 15, 2010.

Offices review the cases and then transfer them to the PGR rather than to local law enforcement agencies, as was previously done. The PGR has agreed to accept any drug smuggling case referred by the U.S. Attorneys, regardless of quality, quantity, or type of illegal drug seized.

Project Gunrunner[177]

Project Gunrunner is an initiative led by ATF in DOJ. Its goal is to disrupt the illegal flow of guns from the United States to Mexico. In addition to its domestic objectives, Project Gunrunner also aims to bolster U.S. and Mexican law enforcement coordination along the border in firearms and violent crime cases as well as to train U.S. and Mexican law enforcement officials to identify firearms traffickers. Project Gunrunner was criticized, in part, for not systematically and consistently sharing information with Mexican and U.S. partners as well as for focusing investigations on gun dealers and straw purchasers over high-level traffickers.[178] In September, 2010, ATF released a new strategy, "Project Gunrunner—A Cartel Focused Strategy," that reportedly sought to addresses those issues.[179]

Electronic Trace Submission System[180]

ATF maintains a foreign attaché in Mexico City to administer an Electronic Trace Submission System (ETSS), also known as the eTrace program, for Mexican law enforcement authorities. In January 2008, ATF announced that e-Trace technology would be deployed to an additional nine U.S. consulates in Mexico (Mérida, Juarez, Monterrey, Nogales, Hermosillo, Guadalajara, Tijuana, Matamoros, and Nueva Laredo).[181] More recently, ATF has developed and deployed a Spanish language version of its eTrace program that enables Mexican authorities to submit firearm trace requests electronically to ATF officials in the United States. From CY2007 through CY2012, Mexican authorities submitted 118,426 firearms recovered in Mexico to ATF for tracing.[182]

Mexican American Liaison and Law Enforcement Training (MALLET)

The FBI created Mexican American Liaison and Law Enforcement Training (MALLET) seminars in 1988.[183] These week-long seminars, hosted at least four times annually in the United States

---

[177] For more information on Project Gunrunner, see archived CRS Report R41206, *The Bureau of Alcohol, Tobacco, Firearms and Explosives (ATF): Budget and Operations for FY2011*, by William J. Krouse.

[178] U.S. Department of Justice, Office of the Inspector General, Review of ATF's Project Gunrunner, I-2011-001, November 2010, pp. iii-v, http://www.justice.gov/oig/reports/ATF/e1101.pdf.

[179] Ibid., p. ix.

[180] For more information on the Electronic Trace Submission System, see archived CRS Report R41206, *The Bureau of Alcohol, Tobacco, Firearms and Explosives (ATF): Budget and Operations for FY2011*, by William J. Krouse.

[181] Bureau of Alcohol, Tobacco, Firearms and Explosives (ATF), Office of Public Affairs, "ATF Expands Efforts to Combat Illegal Flow of Firearms to Mexico," January 16, 2008.

[182] Bureau of Alcohol, Tobacco, Firearms and Explosives, *Data Source: Firearms Tracing System, Calendar Years 2007 - 2011*, March 12, 2012. Data for CY2012 provided to CRS by ATF.

[183] For more information, see Federal Bureau of Investigation, *On the Border: Training Our Mexican Colleagues*, May 11, 2009 http://elpaso.fbi.gov/boader051109.htm.

throughout the four Southwest border states, train Mexican law enforcement officers on various topics including law enforcement management and investigative techniques. The Mexican law enforcement officials participating in these trainings come from all levels of government—federal, state, and municipal. These seminars provide not only training, but opportunities for building trusted partnerships on both sides of the border. The MALLET seminars are funded through the FBI's Office of International Operations.[184]

Policia Internacional Sonora Arizona (PISA)

The Policia Internacional Sonora Arizona (PISA) is a nonprofit organization that was established in 1978 and has continued to enhance international law enforcement communication and train officers in laws and procedures across borders.[185] With nearly 500 representatives from various levels of Mexican and U.S. government, PISA promotes training and mutual assistance to extradite fugitives and solve crimes from auto thefts to homicides. For example, state and local law enforcement from Arizona have been involved in providing tactical, SWAT, and money laundering training to Mexican police.

Author Contact Information

Clare Ribando Seelke
Specialist in Latin American Affairs
cseelke@crs.loc.gov, 7-5229

Kristin Finklea
Specialist in Domestic Security
kfinklea@crs.loc.gov, 7-6259

Acknowledgments

June S. Beittel, Analyst in Latin American Affairs, contributed to this report.

---

[184] From CRS communication with FBI representative, April 27, 2010.

[185] For more information on PISA, see the website at http://www.azpisa.org/.